Advances of Basic Science for Second Generation Bioethanol from Sugarcane

Marcos S. Buckeridge • Amanda P. De Souza
Editors

Advances of Basic Science for Second Generation Bioethanol from Sugarcane

Springer

Editors
Marcos S. Buckeridge
Department of Botany, Laboratory of Plant Physiological Ecology (LAFIECO)
Institute of Biosciences
University of São Paulo
São Paulo, Brazil

Amanda P. De Souza
Department of Botany, Laboratory of Plant Physiological Ecology (LAFIECO)
Institute of Biosciences
University of São Paulo
São Paulo, Brazil

ISBN 978-3-319-49824-9 ISBN 978-3-319-49826-3 (eBook)
DOI 10.1007/978-3-319-49826-3

Library of Congress Control Number: 2017930932

Printed on acid-free paper

This Springer imprint is published by Springer Nature
The registered company is Springer International Publishing AG
The registered company address is: Gewerbestrasse 11, 6330 Cham, Switzerland

Contents

About the Editors

Marcos S. Buckeridge is Associate Professor at the Department of Botany of the University of São Paulo and Director of the National Institute of Science and Technology of Bioethanol (INCT do Bioetanol), which gathers 32 laboratories in 6 states of Brazil with several collaborations in the USA and Europe. Buckeridge published more than 150 papers and chapters and 3 books. From 2009 to 2012 he was Scientific Director of the Brazilian Bioethanol Science and Technology Laboratory (CTBE), in Campinas. He is Reviews Editor for the International Journal *Trees: Structure and Function* (Springer) and Communicating Editor for *Bioenergy Research* (Springer). In 2010, Buckeridge has been appointed as lead author for the next Intergovernmental Panel of Climatic Changes (IPCC) report (AR5) to be released in 2014. He is the President of the Academy of Sciences of the State of São Paulo, Brazil.

Amanda P. De Souza has a Ph.D. in plant physiology and biochemistry and currently works as a postdoctoral researcher at the University of Illinois in Champaign-Urbana. De Souza has 12 years of experience working with physiological responses of crops to global climate changes in Brazil and in the USA. In her young career, she published about 30 papers and chapters. From 2008 to 2015, she was the Scientific Manager of the National Institute of Science and Technology of Bioethanol (INCT do Bioetanol), assisting research coordination in more than 30 laboratories across Brazil. During this period she was also involved in projects between Brazil and Europe related to cell wall composition in bioenergy crops.

Chapter 1
Routes to Second-Generation Bioethanol in Brazil: Foundation of the National Institute of Science and Technology of Bioethanol

Marcos S. Buckeridge and Amanda P. De Souza

Abstract Along the last 8 years, the National Institute of Science and Technology of Bioethanol (INCT-Bioetanol) reached several important milestones towards the main objective of producing second-generation bioethanol from sugarcane bagasse. The sugarcane cell wall structure and architecture is now better understood, dozens of new cell wall hydrolases had their genes and enzymes fully characterized, sugargane physiology and its responses to climate change were investigated, and novel yeast strains capable to metabolize pentoses have been discovered. At the same time, history and politics involved in sugarcane as a bioenergy source have been studied and revised. Thus, simultaneously with other large bioenergy programs in the USA and Europe that were set up to improve renewable energy production, the INCT-Bioetanol in Brazil managed to produce important data in basic research. All this data is expected to help industry to produce not only more ethanol, but also new biomaterials to foster bioeconomy. This book reports some of these findings.

Keywords Bioenergy • Bioethanol • Second generation • FAPESP • BioEn • INCT

The National Institute of Science and Technology of Bioethanol (INCT-Bioethanol) was founded in 2008, at the same time that other international initiatives such as the Energy Biosciences Institute (EBI) in the United States and the European Bioenergy Research Institute (EBRI) in England. The main goal of all those

M.S. Buckeridge (✉)
Laboratory of Plant Physiological Ecology (LAFIECO), Department of Botany, Institute of Biosciences, University of São Paulo, Sao Paulo, Brazil
e-mail: msbuck@usp.br

A.P. De Souza
Laboratory of Plant Physiological Ecology (LAFIECO), Department of Botany, Institute of Biosciences, University of São Paulo, Sao Paulo, Brazil

Carl R. Woese Institute for Genomic Biology, University of Illinois at Champaign-Urbana, Champaign, IL, USA

M.S. Buckeridge, A.P. De Souza (eds.), *Advances of Basic Science for Second Generation Bioethanol from Sugarcane*, DOI 10.1007/978-3-319-49826-3_1

institutes was to contribute to increasing the knowledge and the development of new routes related to bioenergy.

The motivation or the creation of the INCT-Bioethanol was closely related to the launching of the fourth International Panel on Climate Change report (AR4) in 2007, which was strongly highlighted in Brazil. At the same time, the São Paulo Research Foundation (FAPESP) started the Bioenergy Research Program of São Paulo (BIOEN) and a few months latter, the Brazilian Government set up a series of strategic science and technology institutes (the INCT Program) to try to solve problems that could improve the industrial performance in the country. Just after the foundation of BIOEN, a group of researchers gathered in a workshop to discuss the scientific challenges to be faced in order to lead sugarcane bioethanol to a new technological status. After this workshop, 33 laboratories across six different Brazilian states became part of the INCT-Bioethanol proposal.

It was clear that the production of renewable energy could significantly contribute to mitigating CO_2 emissions in the planet, and that Brazil was already contributing to such mitigation due to previous development of consistent technologies to produce bioethanol from sugarcane. At that moment, however, the available technologies to produce bioethanol relied only on the use of sucrose (first-generation bioethanol), whereas the bagasse was being used exclusively to produce electricity and the leaves had no or little use. So, the presentations and discussions during the workshop that originated the INCT-Bioethanol proposal involved questions related to the possibility of use of those residues to significantly increase the production of bioethanol in Brazil and to the knowledge that would be necessary to achieve this goal. There were formidable problems to be solved and the group of researchers present at the meeting quickly realized that it was necessary to start a research project focused on basic science in order to produce the knowledge that could be used for technological development.

Some points highlighted during the 2008 workshop became milestones of INCT-Bioethanol for the decade to come. These milestones were the following:

1. *Understand the structure and architecture of the sugarcane cell walls*: Although the main components of sugarcane cell walls were already known, there was lack of knowledge related to the structure of the polymers (polysaccharides and lignin), and to the glycosidic linkages and interactions among polymers that needed to be explored to help understand how to hydrolyze sugarcane cell walls.
2. *Find ways to hydrolyze sugarcane cell walls*: In 2008, both sulfuric acid and enzymatic hydrolysis were available options to hydrolyze sugarcane bagasse. However, local initiatives such as the Dedini pilot plant that were using sulfuric acid ended up discontinuing its unit, and the acid route was abandoned. Thus, the focus started on the enzymatic routes. But, very little was known about enzymes capable of hydrolyzing biomass. More than that, it was clear that at the same time as the cell wall of sugarcane would become known, it would be possible to engineer enzymes to improve their catalytic properties. In addition, it was realized that modern molecular biology tools could be used to engineer the sugarcane, so that endogenous hydrolytic mechanisms could be controlled and used to help the hydrolysis as a biological pretreatment.

3. *Understand sugarcane photosynthesis and growth, especially under climate change conditions*: We were aware that the biomass-producing capacity of some of the sugarcane varieties was much higher than it was seen in the field. Thus, it was possible to turn sugarcane even more productive than thought at that time. For that, physiological, metabolic, and gene expression sugarcane responses to different stresses would have to be deeply studied, so that we would be able to give support to the production of sugarcane in a changing climate.
4. *Understand sugarcane genome and genetics*: We knew that it would be quite complicated since sugarcane is a polyploid and we needed to develop new tools to manipulate sugarcane genome and genetics. The traditional breeding coupled with the development of those new tools would bring really great advances in this area. Another challenge was to genetically transform sugarcane. As we intended to learn more about gene expression related to several processes, it would be natural to transform plants either for purely scientific purposes (proofs of concepts) or to produce new varieties that could actually be used in the field.
5. *Find ways to improve extant yeast strains or discover novel species of them that could ferment pentoses efficiently and produce ethanol*: Although the Brazilian 1G technology was highly efficient for ethanol production from sucrose, the strains used did not process pentoses with the same efficiency. Thus, genetic manipulation of known strains and screening of microorganisms capable of fermenting pentose were necessary.
6. *Develop new pretreatments and associate them with hydrolysis and fermentation*: Sugarcane biomass, both bagasse and trash, were hard to hydrolyze and new methods to pretreat biomass had to be tested in order to suggest possible procedures to be used in industry.

The main mission of the INCT-Bioethanol was the production of knowledge that could help industry, rather than the development of specific technologies. Other organizations were set up at the same time to take care of the technological development, such as the CTBE (National Laboratory of Science and Technology of Bioethanol) in Campinas, São Paulo. This center was designed at that time to bridge the gap between basic science and technological development of the sector, making the links with industry.

After 8 years of research in this field, Brazil has developed a couple of initiatives to produce second-generation (2G) ethanol from sugarcane, but the integration of the basic knowledge into industrial technology has proven to be rather slow. It has to be noted, though, that the forecasts made by the INCT-Bioethanol were that the basic research needed at least one decade to produce enough knowledge and provide an efficient connection with industry. Therefore, we expect that the knowledge produced by the INCT-Bioethanol will start to be incorporated within the middle of the second decade of the twenty-first century, depending on the economic return that such incorporation signalizes to the companies.

The INCT-Bioethanol has not yet completely solved yet the main problems stated in the six milestones listed above, but it has significantly advanced the science necessary for sugarcane bioethanol technology development. Some of the achievements made and their implications to society are shown in Fig. 1.1. The

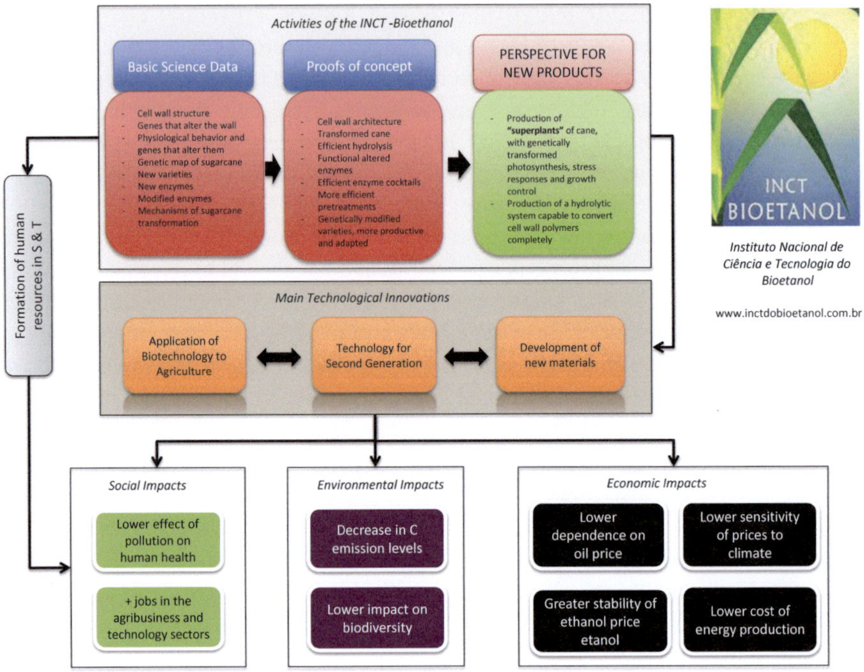

Fig. 1.1 Scheme with the main activities of the INCT-Bioethanol and its consequences for innovation and sectors of the society

science produced by the institute has been reported in hundreds of papers, book chapters, and theses released by the research groups along the last 8 years (http://www.inctdobioetanol.com.br). The first book was published in 2011, reporting the advances reached at that time (Buckeridge and Goldman 2011). We now present a second book, which brings chapters that update aspects related to the sugarcane cell wall (Chaps. 2, 3, and 4) and genetics (Chap. 9), enzymes (Chaps. 5, 6, and 7), pentose fermentation (Chap. 8), sustainability (Chap. 10), and international policies (Chap. 11).

Although this book does not cover all the science produced by the INCT-Bioethanol, it brings an important sample of some of the main achievements reached by the group after almost one decade of hard work. We expect that the scientific knowledge compiled in this book will help students, researchers, and the industry to improve even more the production of 2G bioethanol, influencing industrial and environmental sectors and helping to use the benefits of renewable energy for mitigation and adaptation to the global climate changes.

Reference

Buckeridge MS, Goldman GH (2011) Routes to cellulosic ethanol. Springer, New York, 270 p

Part I
Aspects of Cell Wall Structure and Architecture

Chapter 2
Sugarcane Cell Wall Structure and Degradation: From Monosaccharide Analyses to the Glycomic Code

Marcos S. Buckeridge, Amanda P. De Souza, Eveline Q.P. Tavares, and Arthur B. Cambler

Abstract Pretreatments and enzymes have been a major hindrance to second-generation (2G) bioethanol production. As a result, most scientists have been focusing on the search for new enzymes and their subsequent characterization. Although this valuable knowledge has significantly improved the field generating initiatives of commercial production of 2G bioethanol, the cell walls themselves have received relatively little attention. In this chapter, we revise the work performed on sugarcane cell wall composition, structure, and architecture. From the status of looking exclusively to monosaccharide composition, research has evolved and several details about sugarcane cell wall polysaccharides and lignin were unrevealed. The studies about cell wall structure led to the proposition of the first model of sugarcane cell wall architecture in which macrofibrils (bundles of microfibrils) of cellulose are likely to be bound together by xyloglucan and arabinoxylans. These macrofibrils are covered with layers of more soluble hemicelluloses such as highly branched arabinoxylans and β-glucan. The lignin seems to be closely associated with the cellulose–hemicellulose domain, which is more hydrophobic than the other cell wall domains. Finally, lignin and cellulose–hemicellulose domains are embedded in a thin layer of pectin matrix. This model led to the proposition of a hypothesis that efficient cell wall degradation in the natural environment could be possible if the glycosyl hydrolases would sequentially degrade each layer at a time inwards towards cellulose microfibrils. This hypothesis was corroborated both during the attack of fungi to sugarcane biomass and during the aerenchyma formation in sugarcane

M.S. Buckeridge (✉) • E.Q.P. Tavares • A.B. Cambler
Laboratory of Plant Physiological Ecology (LAFIECO), Department of Botany, Institute of Biosciences, University of São Paulo, Sao Paulo, Brazil
e-mail: msbuck@usp.br

A.P. De Souza
Laboratory of Plant Physiological Ecology (LAFIECO), Department of Botany, Institute of Biosciences, University of São Paulo, Sao Paulo, Brazil

Carl R. Woese Institute for Genomic Biology, University of Illinois at Champaign-Urbana, Champaign, IL, USA

M.S. Buckeridge, A.P. De Souza (eds.), *Advances of Basic Science for Second Generation Bioethanol from Sugarcane*, DOI 10.1007/978-3-319-49826-3_2

roots. The highly complex sugarcane cell wall is now thought to be a result of a code, which is just starting to be unveiled. We believe that by further understanding the interactions among polymers and how endogenous enzymes attack cell walls, future strategies to induce endogenous biological pretreatments followed by the attack of enzyme consortia might significantly improve industrial processes for 2G bioethanol production from sugarcane.

Keywords Cell wall architecture • Polysaccharides • Cellulose • Hemicellulose • Pectins • Lignin • Bioenergy • Glycosyl hydrolases

2.1 Introduction

Sugarcane breeding has promoted the development of varieties that are highly productive. This is one of the main factors that led Brazil to become the second largest world producer of first-generation (1G) bioethanol, which is produced from sucrose stored in culms of sugarcane. More recently, the advances in second-generation (2G) bioethanol science have endorsed the development of at least two Brazilian initiatives for commercial production of 2G bioethanol: GranBio and Raizen/Costa Pinto. Currently, the former utilizes sugarcane leaves as the main raw material for the 2G bioethanol production, while the second uses sugarcane bagasse, integrating 1G and 2G processes.

Second-generation bioethanol became a reality due to the significant progress that has been achieved in the development of processes to pretreat and hydrolyze the plant cell wall (Vohra et al. 2014; Healey et al. 2015; Saini et al. 2015). Despite the great advances, the hydrolysis step, reported to account for 9% of the cost of biomass conversion in 2002 (Aden et al. 2002), remains as a substantial parcel of the cost of 2G processes (Enzitec 2016). One of the factors that can help to reduce the costs in 2G processes is the formulation of more accurate enzymatic cocktails, designed to hydrolyze cell wall polysaccharides efficiently. For this purpose, it is crucial to understand how polysaccharides are structured and arranged into sugarcane cell walls, i.e., its architecture (Buckeridge et al. 2016—Fig. 2.1).

In this chapter, we review some of the work performed on sugarcane cell wall composition and architecture and discuss future scenarios of research to improve the efficiency of cell wall degradation in sugarcane.

2.2 Molecular Composition of Sugarcane Cell Walls

Sugarcane cell wall composition has been firstly estimated by analyzing monosaccharides. Peng et al. (2009) have shown that cell walls from sugarcane bagasse were composed of 38.3% glucose, 27.6% xylose, and 19.2% galactose. Later on, Masarin et al. (2011) reported the presence of 38–43% glucans, 25–32%

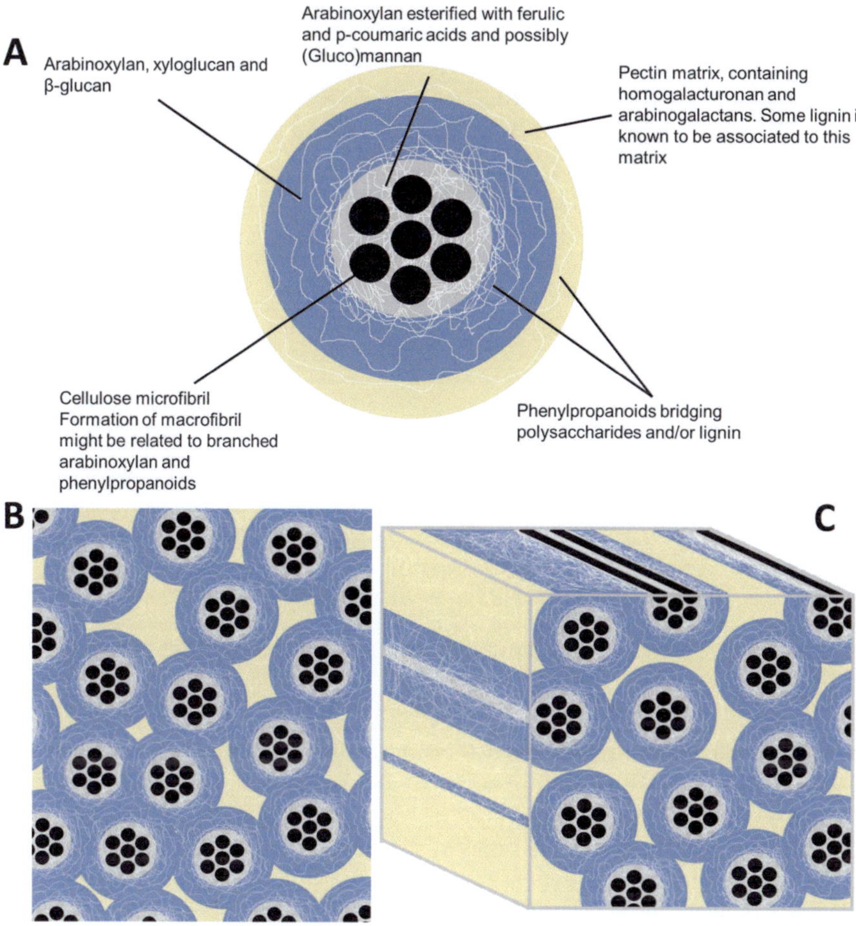

Fig. 2.1 Hypothetical model of the sugarcane cell wall. Based on the information provided by the *cell wall phenotyping* procedures (De Souza et al. 2013; Buckeridge et al. 2015) different classes of polymers are arranged according to their solubility and topology based on antibodies and atomic force microscopy (Buckeridge et al. 2015). Sugarcane walls are of the type II (grasses), with relatively little pectin matrix and having arabinoxylans > β-glucan > xyloglucan > mannan as the main hemicelluloses. (**a**) Representative scheme of what could be the basic repetitive unit of the cell walls of sugarcane (possibly in primary and secondary walls, but with different proportions among domains). (**b, c**) Representative schemes of what could be a tridimensional portion of the primary cell wall of sugarcane. Secondary walls would probably have less of the hemicellulose domain and more cellulose *macro*fibrils. This tentative model does not include cell wall proteins, such as hydroxyproline-rich proteins that have not been studied in sugarcane to date

hemicelluloses, as well as 1.6–7.5% extractives in bagasse of different sugarcane varieties produced in Brazil.

A more accurate appreciation of sugarcane cell wall composition was achieved by De Souza et al. (2013), who, besides monosaccharides, used a combination of

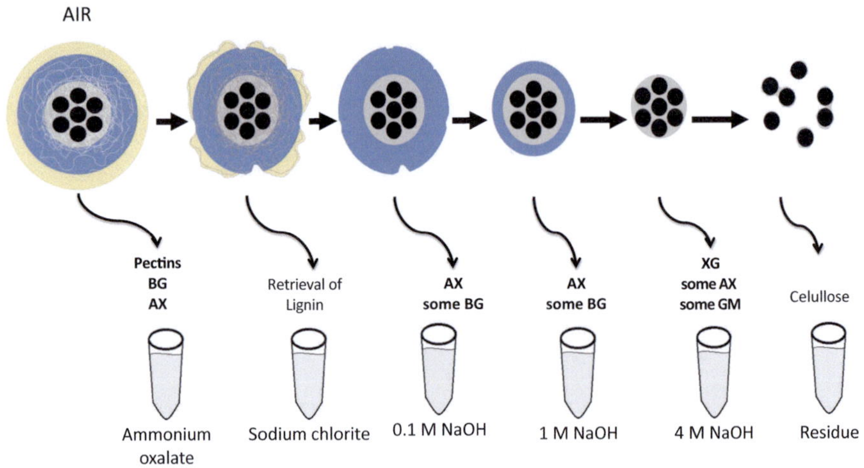

Fig. 2.2 Chemical fractionation of the cell wall as it occurs with sugarcane culm according to De Souza et al. (2013). Alcohol-insoluble residue (AIR) is prepared by washing dry biomass with 80% ethanol at 80 °C. AIR is then submitted to a sequence of extractions. β-Glucan (BG) and pectins are solubilized in ammonium oxalate, and lignin is retrieved from the material by sodium chlorite. A sequence of increasing concentrations of alkali (NaOH) will extract first arabinoxylans (AX) and some BG (0.1 and 1 M) and then xyloglucan (XG) with some arabinoxylan and (gluco)mannan (GM) (4 M). The residue ends up composed of 98% glucose, which is assumed to be all cellulose

different techniques such as oligosaccharide and polysaccharide profiling and Fourier transform infrared spectroscopy (FT-IR). Combined, these procedures can be defined as a *cell wall phenotyping* procedure. In this study, cell walls of leaves and culm were fractioned with a series of solvents so that polymers were extracted according to their solubility (Fig. 2.2). The composition of the polymers in each fraction was analyzed so that it was possible to estimate the average composition of each class of polymer (pectins, hemicelluloses, and cellulose) of the cell wall of this plant. On average, sugarcane cell walls from both organs were shown to be composed of ~30% cellulose; 40% arabinoxylan (AX); 10% β-glucan (BG); 8% xyloglucan (XG); 8% pectins including homogalacturonan, arabinogalactan, and arabinan; and 6% lignin.

Although the data compilation made by De Souza et al. (2013) led to a cell wall composition estimation that was quite different from other studies on bagasse (Rezende et al. 2011; Guilherme et al. 2015), one has to consider that both leaves and culm analyzed were from sugarcane *in natura*. The composition of *in natura* culms differs from bagasse since the latter loses some of the polysaccharides during the extraction of sucrose in the industry (Lara Azevedo, Amanda P. De Souza, and Marcos Buckeridge, unpublished results). The lignin content, though, might have been underestimated due to plant young age, possibly having a little interference in the hemicellulose proportion in the wall.

Some of the sugarcane cell wall polymers have been investigated in detail. For example, the identity of the carbohydrate linkages of the main hemicelluloses

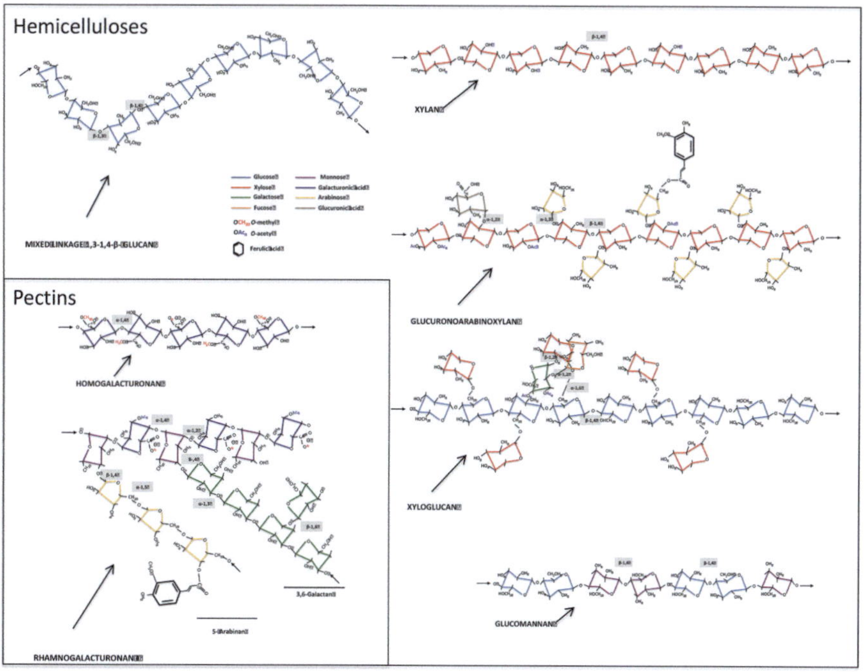

Fig. 2.3 Structural features of the cell wall polysaccharides from sugarcane. Redesigned from Buckeridge and De Souza (2014). Molecular structures for each polysaccharide were based on methylation analysis of sugarcane cell wall fractions (Marcos Buckeridge and Nicolas C. Carpita, unpublished results)

and pectins has been revealed through methylation analysis (see Fig. 2.3 for structures), and the cellulose nanostructure has been explored in depth (see Chap. 3 for further information). Also, the fine structure of the hemicelluloses has been determined by using endo-β-glycanases (De Souza et al., 2013). All of these features are known to contribute to the polysaccharide arrangement, and consequently to the cell wall architecture.

2.3 Sugarcane Cell Wall Architecture

Cell walls can be thought of as a stratified composite formed by well-organized interactions (covalent and non-covalent) among polymers. There seems to exist a basic module composed of cellulose, hemicelluloses, and pectins that is replicated within the wall (Fig. 2.1a), forming the network of compounds that emerge as the cell wall becomes proper (Fig. 2.1b, c).

In sugarcane, the model proposed for the basic module is based on the pattern of solubilization of polysaccharides described by De Souza et al. (2013), and on recent results obtained in our laboratory (Arthur Cambler, Amanda De Souza, and Marcos Buckeridge, unpublished results). The sugarcane basic model (Buckeridge et al. 2016) is formed by cellulose *micro*fibrils that are probably bound together by arabinoxylans (AX) whose branches are esterified with phenylpropanoids (Fig. 2.1). The presence of phenylpropanoids, as well as acetylations, in the AX chains might confer extreme hydrophobic properties to this inner part of the cell wall module. The *macro*fibrils seem to be covered with more branched (and possibly less acetylated) AX that confers relative hydrophobicity to the wall. A small proportion of mannan (or glucomannan) has been detected in the less soluble sugarcane cell wall fractions, suggesting that this polysaccharide might also interact with *micro*fibrils and participate in the complex that forms *macro*fibrils. However, this hypothesis needs further investigation. The most soluble hemicellulosic polymers in the walls are mostly linear, but slightly arabinosylated xylan and the mixed linkage β-glucan (BG). It has been proposed that BG plays a structural role in the wall, mainly by forming a scaffold for deposition of other wall polymers during development (Buckeridge et al. 2004). Alternatively, BG has been found to play a storage reserve role in wheat (Roulin and Feller 2001) and barley (Roulin et al. 2002).

Lignin of sugarcane is mostly associated with vascular bundles (mainly in its fibers) and a little—primarily in rind cells—is located in the walls of the parenchyma (composed mainly of *p*-coumaric acid), followed by significant amounts of hydroxycinnamic and ferulic acids (Masarin et al. 2011) (Fig. 2.4). However, it is not well known how these compounds are chemically linked to each other. What is known is that hydroxycinnamic and ferulic acids are esterified to hemicelluloses and pectins, nucleating the polymerization of lignin in the wall (see Chap. 4). Thus, there seems to exist a trend towards the presence of lignin in cell walls that contain more cellulose, whereas pectins, probably in the middle lamella, also contain some lignification. Recent evidence strongly suggests that AX is the main polymer bound to lignin, probably via its arabinosyl branching residues. Because a portion of the branched AX is retrieved from the walls of sugarcane only after extraction with sodium chlorite (Cambler, De Souza, and Buckeridge, unpublished results), we believe that lignin bridges are more frequent among AX molecules and somehow between them and cellulose.

Although chemistry has helped understanding some aspects of the interactions among cell wall polymers, their topology within the wall is not yet accessible by any technique. The application of modern techniques to unveil cell wall features related to the physics, chemistry, and biochemistry of the cell wall of sugarcane has led to a working model of its architecture. However, we still lack tools that can show how the polymers are arranged in the native wall. Those tools are essential to advance the knowledge of the cell wall architecture. For that purpose, probes such as DNA aptamers could be developed to provide identification of polymer domains (Boese and Breaker 2007; Low et al. 2009). Nonetheless, the barriers to be crossed are enormous. For instance, the use of such probes, which could bypass the problem of the low wall porosity, would still face the hydrophobicity barrier of the deep regions of the wall, representing some technical issues that would be quite complex to deal with.

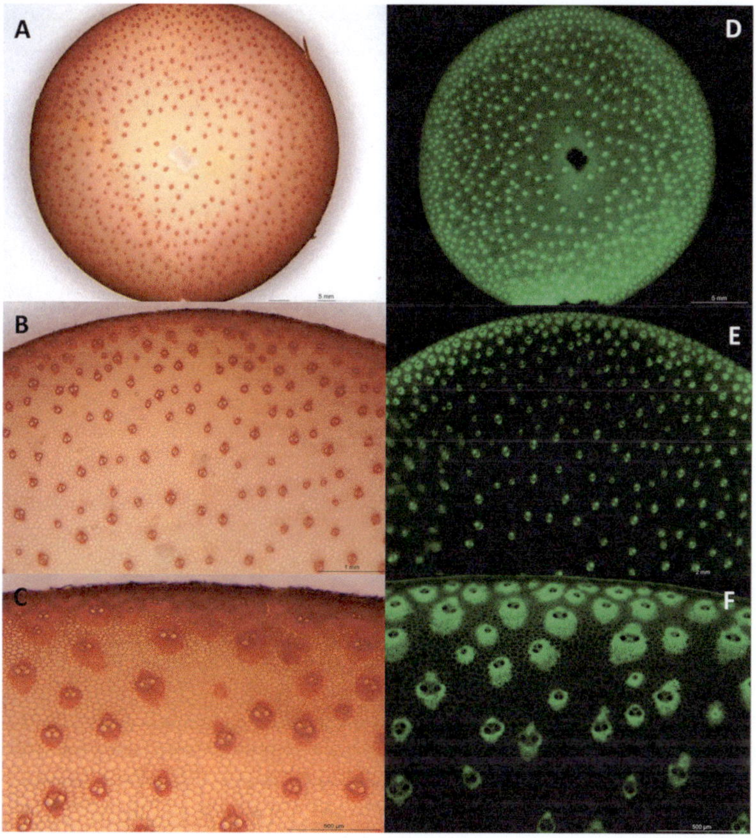

Fig. 2.4 Distribution of lignin in the culm of sugarcane. Lignin was stained with floroglucinol (**a–c**) and visualized by autofluorescence (**d–f**). Section of the culm was obtained from mature culm of sugarcane (cv. SP80-3280). Lignin is more concentrated in the vascular system, especially in the fibers (see mainly **c**), but is also detected in parenchyma cell walls, although staining less strongly in these cells. There is proportionally higher lignin concentration in the periphery of the culm due to higher incidence of vascular bundles. Pictures taken by Viviane C. Lopes

2.4 The Cell Wall Architecture Results from a Glycomic Code

From the discovery of the grass wall architecture to the modern days, a plethora of genes related to plant cell wall biosynthesis and hydrolysis have been described and characterized (see Wang et al. 2016 for a review). The knowledge about mechanisms of synthesis and hydrolysis of the cell walls in general highlights the fact that carbohydrate polymers are not composed of randomly linked monosaccharides, but by polymers that display very strict fine structures (Buckeridge and De Souza 2014). The fact that encoded polymers form such a complex structure raised the question whether cell walls could have a code that would be encrypted by the biosynthetic mechanisms of different classes of polymers (Buckeridge and De Souza 2014; Tavares and Buckeridge 2015).

The idea that cell wall polymers may display a code—named the glycomic code—has been put forward to explain how seed storage cell wall polysaccharides are degraded (Buckeridge 2010). In this case, XG clearly displays encoded fine structure that determines the action of glycosyl hydrolases (Tiné et al. 2003, 2006) as well as the level of binding to the surface of cellulose (Lima and Buckeridge 2001).

In the case of sugarcane cell walls, AX, the major component among hemicelluloses, seems to be encrypted by branching with arabinosyl residues, which are positioned at carbons 2 and/or 3 on xylosyl residues in the main chain. Many of the unbranched hydroxyls are supposedly acetylated making most of the polysaccharide hardly accessible to enzymes (Crivellari 2012).

The glycomic code may be the ultimate barrier to efficient hydrolysis. If so, by breaking the glycomic code of all cells in a tissue, one could supposedly control cell wall assembly and, by knowing precisely the mapping of polymer interactions, be able to hydrolyze cell walls with much higher precision.

2.5 What Structure and Architecture Mean for Hydrolysis

The level of complexity of the sugarcane cell walls is not surprising since some features of the architecture of cell walls of grasses have been known for decades (Carpita and Gibeaut 1993). However, the idea of hydrolyzing its components for bioenergy purposes remains a major challenge. Part of this challenge can be attributed to the lack of knowledge about cell wall architecture.

One of the main features of the cell wall architecture relies on the fine structure of the polysaccharides, as it determines how the polymers can be arranged in the wall. The study of the fine structure can be made using endo-hydrolases. When in low concentration, the endo-enzyme action is analogous to a DNA restriction enzyme. Thus, the use of endo-enzymes such as lichenase for BG, xyloglucan-endo-glucanase for XG, and GH10 or GH11 for AX produces "restriction maps" of oligosaccharides that can be viewed in HPAEC-PAD chromatograms. When such maps are obtained for cell wall fractions, they might reveal differences in polysaccharide fine structure that may be relevant for hydrolysis (De Souza et al., 2013, 2015). For instance, AX from sugarcane leaves is more branched with arabinose and seems to bind more strongly to cellulose in comparison to culm; that is, the AX from leaves remains bound up to the 4 M NaOH fraction during fractionation. On the other hand, XG seems to have more soluble fractions in culm than in leaves (Fig. 2.5), which may reflect differences in saccharification between the two organs since xyloglucan may interfere with the attack of enzymes to cellulose.

Differences in the glycome profiling, i.e., identification of polysaccharide epitopes through monoclonal antibodies (Patthathil et al. 2012), can also suggest alterations in the fine structure since they give information about changes in the populations of the exposed epitopes present in the cell wall fractions (Zhu et al. 2010).

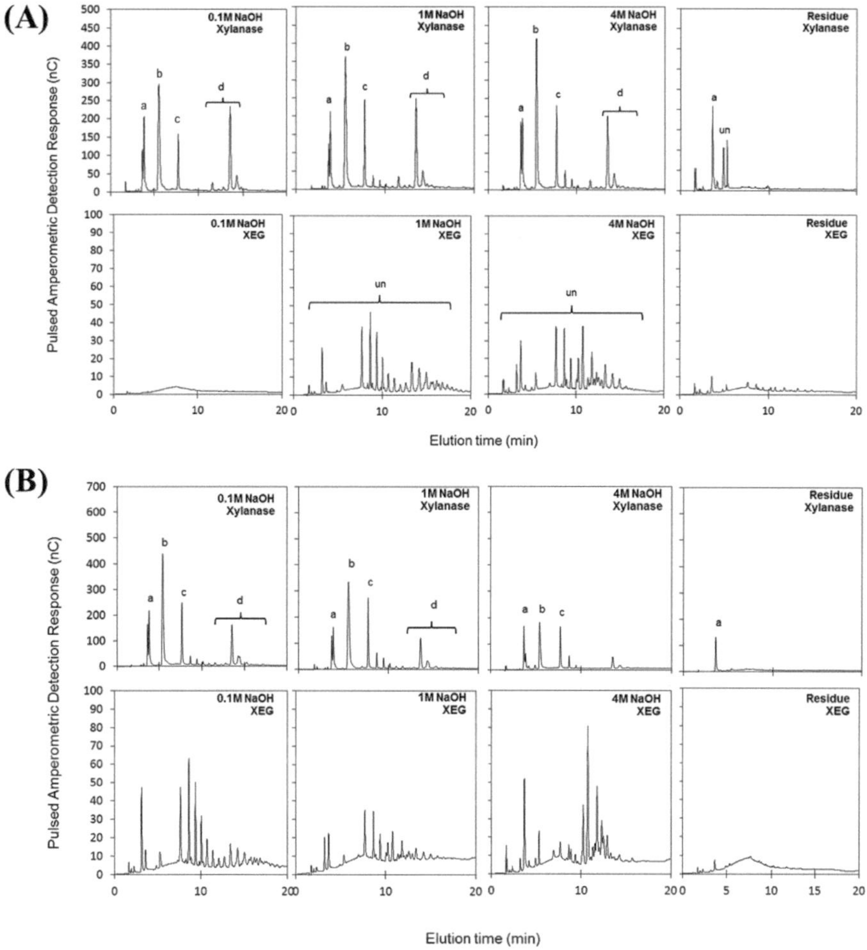

Fig. 2.5 Comparison of the fine structures (restriction profiles) of the main hemicelluloses of sugarcane leaves (**a**) and culm (**b**). Adapted from De Souza et al. (2013). Xylanase used for detection of AX oligosaccharides and xyloglucan endo-glucanase (XEG) for detection of XG oligosaccharides. *a* xylose, *b* xylobiose, *c* xylotriose, *d* arabinoxylated oligosaccharides, *un* unknown xyloglucan oligosaccharides

In miscanthus, a closely related species to sugarcane, subtle differences in pectins were identified by glycome profiling such as the presence of arabinogalactans that were relevant to saccharification (De Souza et al. 2015).

The arrangement of the polymers into an architectural framework also raises issues related to pore size in the biomass, which apparently limits the penetration of enzymes on it (Buckeridge et al. 2016). Based on the polysaccharide structure and composition, we assume that at least 24 distinct linkages would require enzymes to be broken. For this hydrolysis process, at least 18 classes of enzymes would be needed (Table 2.1). Thus, to overcome the limitation of pore size and have an efficient

Table 2.1 Estimates of the average composition of sugarcane cell walls on the basis of the data produced from leaves and culm of sugarcane by De Souza et al. (2013)

Class (%)	Polymer (%)	Chemical composition (number of carbons in parenthesis)	Hydrolases
Cellulose (28%)	Microfibrils of glucans (28%)	Glucose (6)	Cellulases, β-glucosidases
Hemicelluloses (58%)	Arabinoxylan (40%)	Xylose (5), arabinose (5), ferulic acid (10), acetyl (2)	Endo-xylanase, arabinofuranosidase, feruloyl-esterase, β-xylosidase
	β-Glucan (10%)	Glucose (6)	Lichenase, β-glucosidase
	Xyloglucan (8%)	Glucose (6), xylose (5), galactose (6), ferulic acid (10), acetyl (2)	Xyloglucan endoglucanase, β-galactosidase, α-xylosidase, feruloyl-esterase
Pectins (8%)	Homogalacturonan	Galacturonic acid (6), methyl (1)	Endopolygalacturonase, pectin methyl esterase
	Arabinogalactan	Arabinose (5), galactose (6)	Arabinosidases, galactanases
	Arabinnan	Arabinose (5)	Arabinosidase, arabinanases
Lignin (6%)	Polymers of phenylpropanoids	Guaiacyl (7), syringyl (11), conyferil (10)	Laccases

The monosaccharide composition and the possible enzymes that attack each polymer of the wall. Reprinted from De Souza et al. (2014)

hydrolysis, the basic architectural unit (Fig. 2.1a) would have to be attacked by enzymes acting from the surface towards the inner side of the *macro*fibril. Following this idea, De Souza et al. (2013) proposed a hypothetical mechanism by which a group of enzymes in a cocktail would have their action in a sequence of attacks. This attack would start with esterases (pectin methyl and acetyl esterases, xylan acetyl esterases, feruloyl esterases), followed by the action of endo- and exo-hydrolases that would attack the hemicelluloses within the wall (AX, BG, XG, and mannan). After such attack, cellulose *micro*fibrils would end up naked and could then be attacked by lytic oxidases, endo-glucanases, and cellobiohydrolases.

This hypothesis received some experimental support by the experiment reported by Borin et al. (2015). These authors used proteomics and biochemistry to show that the production of enzymes in the forecasted sequence took place when *Trichoderma reesei* and *Aspergillus niger* were grown on sugarcane culm and bagasse. A similar sequence of enzyme attacks was also observed in an endogenous cell wall degradation during aerenchyma formation in sugarcane roots (refer to Grandis et al. 2014 and Tavares et al. 2015 for further information regarding endogenous mechanisms in plants that include cell wall degradation).

Together, these discoveries led us to propose that it could be more appropriate to use three enzyme consortia in which synergic properties are taken into consideration. The first consortium would target the most soluble polymers of the wall (pectins and BG debranching enzymes). The second one should be capable of degrading the main hemicelluloses that are bound to the surface of *macro*fibrils (endo- and exo-xylanases, xylosidases, mannanase, and lichenase in certain cases). Finally, the third consortium would contain mostly expansin, cellulose, lytic oxidases, and xyloglucan-degrading enzymes, possibly with some endo-xylanase as well (see Chap. 5 for details about each enzyme). If the consortia were added sequentially, mimicking how the hydrolysis happens in vivo, perhaps much less enzyme would have to be used in the process. The use of lower amounts of enzymes could be achieved, since the enzymes that degrade cellulose, for example, would be freshly added to the mixture instead of being present in the process for several hours without being able to act on their substrates.

The discovery that natural degradation mechanisms display sequential action of glycosyl hydrolases on biomass can also help to design biological pretreatments. One could design plants capable of activating certain cell wall-degrading enzymes just before harvesting, thus facilitating the pretreatment step. Altogether, this could have a potential to decrease significantly the higher costs associated to pretreatment and hydrolysis in the 2G processes.

2.6 Conclusions and Perspectives

In spite of the significant advances in sugarcane biomass hydrolysis, the difficulty to access the high level of complexity of the cell walls clearly shows that there is much more to be studied to unveil hidden details of polysaccharide structures as well as the ways in which they interact within the wall, giving rise to architectural features. The fine structures of sugarcane AX and XG, for instance, remain unknown. The fact that it so happens with the fine structural details of pectin branching is important since we know that pectins play important roles in recalcitrance (Latarullo et al. 2016).

The discovery that a sequential action of glycosyl hydrolases occurs in vivo brings a new perspective to how efficient cell wall dismantling and polysaccharide hydrolysis could be achieved. By imitating natural processes, but at the same time speeding them up, it is possible that biomass could be degraded faster and at the same time with lower enzyme concentrations. By understanding the mechanisms of synthesis and also the forces involved in the assembly of the cell wall composites, one would be able to hydrolyze them efficiently. Furthermore, it would be possible to gain control on how polymers assemble within the cell wall, so that it would be useful not only for bioenergy production but also to design new materials with much higher aggregated value for industry.

References

Aden A, Ruth M, Ibsen K, Jechura J, Neeves K, Sheehan J, Wallace B, Montague L, Slayton A, Lukas J. 2002. Lignocellulosic biomass to ethanol process design and economics utilizing co-current dilute acid prehydrolysis and enzymatic hydrolysis for corn stover. NREL Technical Report, TP-510-32438

Boese BJ, Breaker RR (2007) In vitro selection and characterization of cellulose-binding DNA aptamers. Nucleic Acids Res 35(19):6378–6388

Borin GP, Sanchez CC, De Souza AP, Santana ES, Souza AT, Leme AFP, Squina FM, Buckeridge M, Goldman GH, Oliveira JVC (2015) Comparative secretome analysis of *Trichoderma reesei* and *Aspergillus niger* during growth on sugarcane biomass. PLoS One 10(6):e0129275

Buckeridge MS, Santos WD, Tiné MS, De Souza AP (2015) The cell wall architecture of sugarcane and its implications to cell wall recalcitrance. In: Lam E, Carrer H, Silva JA (eds) Compendium of bioenergy Plants: sugarcane. CRC Press, 125p

Buckeridge MS, dos Santos WD, Tiné MAS, De Souza AP (2016) The cell wall architecture of sugarcane and its implications to cell wall recalcitrance. In: Lam E, Carrer H, da Silva JA, Kole C (eds) Compendium of bioenergy plants: sugarcane. CRC Press—Taylor and Francis Group, Boca Raton, pp 31–50

Buckeridge MS, Rayon C, Urbanowicz B, Tiné MAS, Carpita NC (2004) Mixed linkage (1-3),(1-4)-beta-D-glucans of grasses. Cereal Chem 81(1):115–127

Buckeridge MS, De Souza AP (2014) Breaking the "glycomic code" of cell wall polysaccharides may improve second-generation bioenergy production from biomass. Bioenergy Res 7:1065–1073

Buckeridge MS (2010) Seed cell wall storage polysaccharides: models to understand cell wall biosynthesis and degradation. Plant Physiol 154:1017–1023

Carpita NC, Gibeaut DM (1993) Structural models of primary cell walls in flowering plants: consistency of molecular structure with the physical properties of the cell wall during growth. Plant J 3:1–30

Crivellari AC (2012) Caracterização estrutural das hemiceluloses de paredes celulares de cana-de-açúcar. Dissertação de Mestrado, Instituto de Biociências, Universidade de São Paulo, São Paulo. Pdf download: http://www.teses.usp.br/teses/disponiveis/41/41132/tde-10102012-084959/. Accessed 23 Aug 2016

De Souza AP, Grandis A, Leite DCC, Buckeridge MS (2014) Sugarcane as a bioenergy source: history, performance, and perspectives for second-generation bioethanol. Bioenergy Res 7:24–35

De Souza AP, Kamei CLA, Torres AF, Pattathil S, Hahn MG, Trindade LM, Buckeridge MS (2015) How cell wall complexity influences saccharification efficiency in *Miscanthus sinensis*. J Exp Bot. doi:10.1093/jxb/erv183

De Souza AP, Leite DCC, Pathatil S, MG H, MS B (2013) Composition and structure of sugarcane cell wall polysaccharides: implications for second-generation bioethanol production. Bioenergy Res 6:564–579

De Lima DU, Buckeridge MS (2001) Interaction between cellulose and storage xyloglucans: the influence of degree of galactosylation. Carbohydr Polym 46(2):157–163

Enzitec (2016) Anais do XII Seminário Brasileiro de Tecnologia Enzimática. http://www.ucs.br/site/eventos/enzitec-2016/anais/. Accessed 8 Aug 2016

Grandis A, De Souza AP, Tavares EQP, Buckeridge MS (2014) Using natural plant cell wall degradation mechanisms to improve second generation bioethanol. In: McCann M, Buckeridge MS, Carpita NC (eds) Plants & bioenergy. Springer, New York, pp 211–230

Guilherme AA, Dantas PVF, Santos ES, Fernandes FAN, Macedo GR (2015) Evaluation of composition, characterization and enzymatic hydrolysis of pretreated sugarcane bagasse. Brazil J Chem Eng 32(1):23–33

Healey AL, Lee DJ, Furtado A, Simmons BA, Henry RJ (2015) Efficient eucalypt cell wall deconstruction and conversion for sustainable lignocellulosic biofuels. Front Bioeng Biotechnol 3:190

Latarullo MB, Tavares EQ, Maldonado GP, Leite DC, Buckeridge MS (2016) Pectins, endopolygalacturonases and bioenergy. Front Plant Sci 7:1401

Low SY, Hill JE, Peccia J (2009) DNA aptamers bind specifically and selectively to (1→3)-β-D-glucans. Biochem Biophys Res Commun 378(4):701–705

Masarin F, Gurpilhares DB, Bafa DCF, Barbosa MHP, Carvalho W, Ferraz A et al (2011) Chemical composition and enzymatic digestibility of sugarcane clones selected for varied lignin content. Biotechnol Biofuels 4:55

Patthathil S, Avci U, Miller JS, Hahn MG (2012) Immunological approaches to plant cell wall and biomass characterization: glycome profiling. In: ME H (ed) Biomass conversion: methods and protocols. Springer, New York, pp 61–72

Peng F, Ren JL, Xu F, Bian J, Peng P, Sun RC (2009) Comparative study of hemicelluloses obtained by graded ethanol precipitation of sugarcane bagasse. J Agric Food Chem 57:6305–6317

Rezende CA, de Lima MA, Maziero P, de Azevedo ER, Garcia W, Polikarpov I (2011) Chemical and morphological characterization of sugarcane bagasse submitted to a delignification process for enhanced enzymatic digestibility. Biotechnol Biofuels 4:54

Roulin S, Feller U (2001) Reversible accumulation of (1,3, 1,4)-β-D-glucan endohydrolase in wheat leaves under sugar depletion. J Exp Bot 52(365):2323–2332

Roulin S, Buchala AJ, Fincher GB (2002) Induction of (1,3, 1,4)-beta-D-glucan hydrolases in leaves of dark-incubated barley seedlings. Planta 215(1):51–59

Saini JK, Saini R, Tewari L (2015) Lignocellulosic agriculture wastes as biomass feedstocks for second-generation bioethanol production: concepts and recent developments. 3 Biotech 5(4):337–353

Tavares EQP, De Souza AP, Buckeridge MS (2015) How endogenous plant cell wall degradation mechanisms can help achieve higher efficiency in saccharification of biomass. J Exp Bot 66:4133–4143

Tiné MAS, Lima DU, Buckeridge MS (2003) Galactose branching modulates the action of cellulase on seed storage xyloglucans. Carbohydr Polym 52:135–141

Tiné MAS, Silva CO, Lima DU, Carpita NC, Buckeridge MS (2006) Fine structure of a mixed-oligomer storage xyloglucan from seeds of *Hymenaea courbaril*. Carbohydr Polym 66:444–454

Vohra M, Manwar J, Manmode R, Padgilwar S, Patil S (2014) Bioethanol production: feedstock and current technologies. J Environ Chem Eng 2:573–584

Wang Y, Fan C, Hu H, Li Y, Sun D, Wang Y, Peng L (2016) Genetic modification of plant cell walls to enhance biomass yield and biofuel production in bioenergy crops. Biotechnol Adv 34:997–1017

Zhu X, Patthatil S, Mazumber K, Brehm A, Hahn MG, Dinesh-Kumar SP, Joshi CP (2010) Virus-induced silencing offers a functional platform for studying plant cell wall formation. Mol Plant 3:818–833

Chapter 3
Nanostructure of Lignocellulose and Its Importance for Biomass Conversion into Chemicals and Biofuels

Carlos Driemeier

Abstract Lignocellulosic biomass is a vast renewable resource made by nature with a hierarchical structure going from entire plants down to simple molecules. Part of this structural hierarchy stands at the nanoscale, where cellulose crystals, crystal aggregates, and cell wall lamellas are distinguishable nanostructural elements. This chapter provides an overview of lignocellulose nanostructure, discussing the fundamentals, changes promoted by thermochemical treatments, relevance for enzymatic digestibility, and specificities of sugarcane. With proper consideration of nanostructural features, more rational and efficient deconstruction of lignocellulose can be devised for the purpose of conversion into chemicals and biofuels.

Keywords Cellulose • Lignocellulose • Pretreatment • Bioethanol

3.1 Introduction

By photosynthesis, higher plants convert sunlight into chemical energy. A large fraction of this chemical energy is stored in plant cell walls, which are structural elements designed by nature to withstand mechanical forces and external biological attacks so as to preserve plant structural integrity. Cell wall biomass is the most prevalent type of biomass on Earth, a resource with the potential to replace a significant share of materials, chemicals, and liquid fuels that global society currently obtains from petroleum. In industrial terminology, this type of biomass is usually termed *lignocellulosic biomass* or simply *lignocellulose*, encompassing wood, agricultural residues, and biomass from high-yield energy crops.

C. Driemeier (✉)
Laboratório Nacional de Ciência e Tecnologia do Bioetanol—CTBE, Centro Nacional de Pesquisa em Energia e Materiais—CNPEM, Caixa Postal 6192, 13084-970 Campinas, São Paulo, Brazil
e-mail: carlos.driemeier@bioetanol.org.br

M.S. Buckeridge, A.P. De Souza (eds.), *Advances of Basic Science for Second Generation Bioethanol from Sugarcane*, DOI 10.1007/978-3-319-49826-3_3

Lignocelluloses are inherently nanostructured solids. The conversion of lignocellulose into chemicals and liquid biofuels requires the deconstruction of this nanostructure. The natural resistance for such deconstruction has been termed *recalcitrance*. At this point, production of chemicals and biofuels differs qualitatively from the well-established processes for the production of cellulosic materials such as pulp, paper, cardboard, and, more recently, nanocelluloses. These value-added materials partly preserve the nanostructural features from their lignocellulosic feedstocks. On the other hand, chemicals and biofuels consist of simple molecules; that is, the original nanostructure of the feedstocks is lost during the biomass conversion processes. Importantly, the new scientific challenges for conversion of lignocellulose come at a time when science has multiple tools to investigate nanostructures. Therefore, it should be possible to take advantage of an understanding of lignocellulose nanostructure to devise more rational and efficient deconstruction of lignocellulose.

This chapter provides an overview of lignocellulose nanostructure: how it is formed, how it affects and is affected by processing, and how it may vary among biological materials. General principles and nanostructural features specific to sugarcane are highlighted.

3.2 Composition of Lignocellulosic Biomass

Lignocelluloses are composites of three major macromolecular components: cellulose, hemicelluloses, and lignin. Although this chapter is dedicated to supramolecular nanostructures, for the sake of completeness this section provides a brief introduction to the main molecular entities. *Cellulose* molecules are long linear chains of β-1,4-linked glucopyranosyl units (Albersheim et al. 2010). Although from the molecular perspective cellulose is the simplest lignocellulosic macromolecule, from the structural perspective cellulose is primarily responsible for the formation of hierarchical nanostructure (O'Sullivan 1997; Klemm et al. 2005; Fratzl and Weinkamer 2007). *Hemicelluloses* are a group of cell wall structural polysaccharides, consisting of polysaccharide backbones usually with side groups such as additional sugars, uronic acids, and acetyls. Types and amounts of hemicelluloses vary depending on the type of plant material (Albersheim et al. 2010; Scheller and Ulvskov 2010; De Souza et al. 2013; also see Chap. 2). *Lignin* is a complex phenolic macromolecule made by linking monolignols with a diversity of possible bond types. Amounts of lignin, relative proportion of monolignols, and types of linkages vary with type of plant material (Albersheim et al. 2010; Akin 2008; Henriksson 2009). In addition to these three major macromolecular components, there are minor but potentially important amounts of *pectins* (Xiao and Anderson 2013; De Souza et al. 2013). Moreover, *water* plays an active, pivotal role in nanostructure, especially if one considers that most processing for chemicals and biofuels is performed in aqueous media. Therefore, the structural roles of water should be carefully considered.

3.3 Cellulose Crystals

Cellulose is the major component of lignocellulose and it has the tendency to form periodic molecular arrangements, *i.e.*, *crystals* (sometimes called *crystallites*, a synonym of *small crystals*). Indeed, cellulose can form different types of crystals called allomorphs or polymorphs (O'Sullivan 1997). Each polymorph is a distinct phase in the thermodynamic sense. *Native cellulose* is the term for cellulose that has not been changed to a different form since its synthesis by a given living organism. Native cellulose has the crystal polymorph cellulose I. Cellulose I can be converted to cellulose II by either mercerization (in concentrated aqueous sodium hydroxide) or regeneration (dissolution followed by precipitation). Treatments in liquid ammonia or in some amines can generate cellulose III_I or III_{II}, where the subscripts indicate whether the polymorph originated from cellulose I or II. Heat can further convert celluloses III_I and III_{II} into celluloses IV_I and IV_{II}, respectively (Wada et al. 2006). Nevertheless, more recent evidence (Wada et al. 2004b) suggests that cellulose IV_I is merely cellulose I with a higher degree of crystal disorder. Thus, cellulose IV_I would not be a distinct polymorph. Cellulose I can also be reversibly converted into a high-temperature phase, with the phase transition observed at 220–230 °C in tunicin cellulose (Wada 2002) and at 180 °C in wood (Hori and Wada 2005). In addition to the aforementioned polymorphs of pure cellulose, there are polymorphs containing additional elements, such as ammonia cellulose (Wada et al. 2006), alkali celluloses (O'Sullivan 1997), and cellulose hydrates (Kobayashi et al. 2012).

Because of its abundance in nature, cellulose I is the most studied polymorph. Solid-state ^{13}C nuclear magnetic resonance (NMR) data showed that cellulose I is a composite of two forms called Iα and Iβ, the latter being more stable (Horii et al. 1987; Debzi et al. 1991) and predominant in higher plants (Atalla and VanderHart 1984, 1999). Using high-quality crystals, respectively, from an alga and a tunicate, modern crystal structures of Iα and Iβ celluloses were determined by diffraction crystallography (Nishiyama et al. 2002, 2003a). With such methods, crystal structures of celluloses II, III_I, and III_{II} were also determined (Langan et al. 2001; Wada et al. 2004a, 2009). These modern crystal structures are relatively recent events in the long history of tentative structures for cellulose polymorphs. Conventions for indexing the crystal planes and associated diffraction peaks have coevolved along this history, but adherence to modern conventions is still incomplete in the cellulose research community (French 2014).

Crystal structures are landmarks in cellulose understanding because they provide atomic coordinates within a crystal's tridimensional repeated unit, the so-called unit cell. A remarkable feature from Iα and Iβ crystal structures is that both consist of stacks of molecular layers, with intra- and intermolecular hydrogen bonds (O–H … O) formed within each layer, whereas layer stacking is mediated only by relatively weak interlayer forces (hydrophobic or van der Waals attractions and C–H … O hydrogen bonds) (Nishiyama et al. 2002, 2003a; Jarvis 2003). The essential difference between Iα and Iβ crystal structures is the pattern of relative displacement between adjacent molecular layers (Nishiyama et al. 2003a; Jarvis 2003).

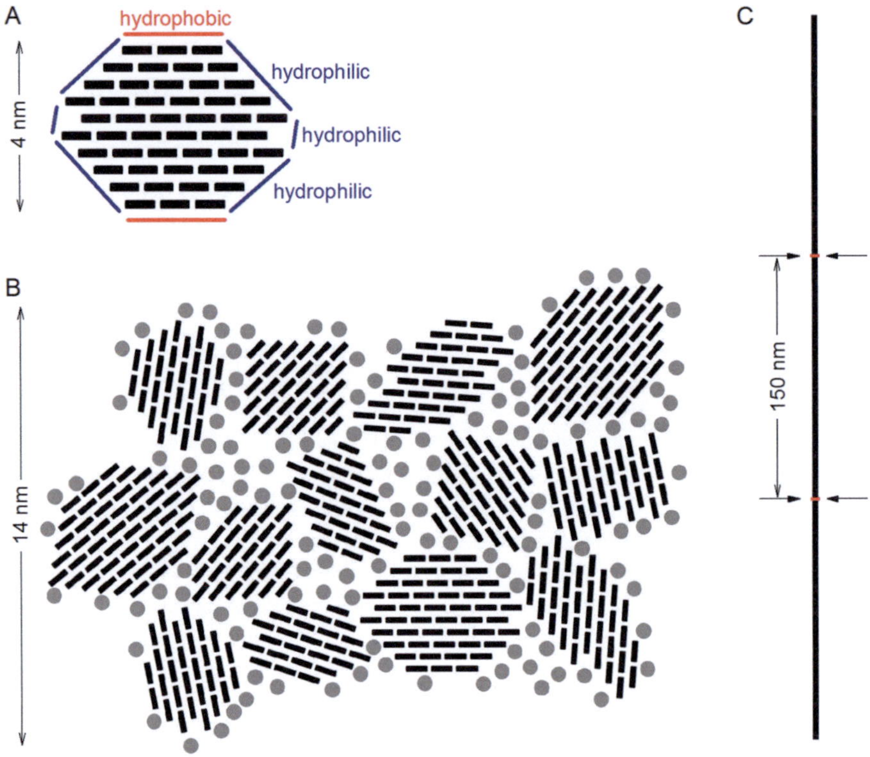

Fig. 3.1 Representation of selected lignocellulose nanostructural elements. (**a**) Cross section of cellulose crystal (cellulose I polymorph). Each *black rectangle* represents the cross section of one cellulose chain. Hydrophobic (*red*) and hydrophilic (*blue*) crystal facets are identified. (**b**) Cross section of cellulose crystal aggregate. *Gray circles* represent non-crystalline hydrated polysaccharides, including hemicelluloses and possibly cellulose itself. (**c**) Longitudinal representation of cellulose fibril formed by long crystalline segments (*black*) spaced by much shorter non-crystalline domains (*red areas, pointed by arrows*). (**a–c**) Characteristic nanometric dimensions are shown for a better sense of scale

Concerning Iα-Iβ coexistence in cellulose from higher plants, a recent study proposed that Iα and Iβ stacking patterns coexist within each crystal, forming a single phase made of crystals with defects known as stacking faults (Driemeier and Francisco 2014).

From Iα and Iβ crystal structures, one also infers that surfaces of cellulose I crystals can have hydrophilic as well as hydrophobic facets (Mazeau 2011) as illustrated in Fig. 3.1a. This distinction is critical for cellulose deconstruction. Hydrophobic and hydrophilic facets differ qualitatively concerning their interactions with lignin, hemicelluloses, and adjacent cellulose crystals (Silveira et al. 2013). Furthermore, the hydrophobic surfaces of cellulose are the sites of preferential adsorption of the carbohydrate-binding module from cellobiohydrolase enzymes (Lehtiö et al. 2003). Moreover, the energy required for chain detachment from crystal surface depends on the chain location in crystal cross section (Beckham et al. 2011).

3.4 Levels of Hierarchical Nanostructure

Nanostructure of lignocelluloses presents a hierarchical organization that results largely from the manner in which lignocelluloses are formed. In higher plants, cellulose is synthesized by hexameric protein complexes known as rosettes (Cosgrove 2005; Somerville 2006). Cellulose chains synthesized simultaneously within one rosette are thought to crystallize (forming cellulose I crystals) into long (up to several micrometers) filaments that some authors (Ding et al. 2014), and this chapter, define as the *cellulose elementary fibril*. "*Elementary*" denotes a fundamental building block, an appropriate term considering the *defined* one-to-one relationship between an elementary fibril and its synthesizing rosette. Noteworthy, the closely related term *microfibril* has deep historical roots and is often employed (Nishiyama 2009), but it is avoided here because we consider the term less precise. The number of cellulose chains in each elementary fibril has provoked considerable debate, with estimates ranging from 18 to 36 (Ding et al. 2014; Fernandes et al. 2011; Newman et al. 2013). From the cellulose Iβ crystal structure (Nishiyama et al. 2002) one knows that each cellulose chain occupies a crystal cross-section area of 0.317 nm^2. Hence, 18–36 chains form crystals with lateral dimensions of 2.4–3.4 nm [calculated as $(18 \times 0.317\ nm^2)^{1/2}$ to $(36 \times 0.317\ nm^2)^{1/2}$]. This range of crystal lateral dimensions is broadly consistent with NMR and X-ray diffraction (XRD) measurements in many (but not all) types of plant cell wall biomass (Fernandes et al. 2011; Newman et al. 2013; Andersson et al. 2003; Driemeier et al. 2012), as we will further detail later.

Nanoscale imaging techniques have enabled visualization of cell wall fibrillar bodies, evidencing widths of a few nanometers. Such fibrillar bodies are here termed *fibrils*. However, correspondence between visualized fibrils and *elementary fibrils* is, in general, not assured. Indeed, nanoscale imaging has shown extensive bundling of cell wall fibrils (Ding and Himmel 2006; Fahlén and Salmén 2005; Zhang et al. 2014), forming fibrillar *aggregates*, which some authors prefer to call macrofibrils (Buckeridge et al. 2016). Such *aggregation* seems to be a fundamental phenomenon in creation of lignocellulose hierarchical nanostructure. Aggregates have been reported to have lateral dimensions of 10–30 nm (Ding et al. 2014; Fahlén and Salmén 2005; Donaldson 2007), which is a few times greater than lateral dimensions of elementary fibrils or, in an alternative reading, a few times greater than lateral dimensions of cellulose crystals. Thus, aggregate cross sections contain *multiple* cellulose crystals (Ding and Himmel 2006; Fahlén and Salmén 2005; Driemeier and Bragatto 2013; Bergenstråhle et al. 2008) as represented in Fig. 3.1b. In general, aggregated crystals have their long axes (i.e., cellulose chain axes) approximately in parallel, while each crystal may have distinct rotations in the cross-sectional plane, implying some crystallographic mismatch between adjacent crystals. In addition, other cell wall polysaccharides such as hemicelluloses and pectins associate with cellulose crystal surfaces, possibly outside as well as inside of the aggregates, so that such polysaccharides act as spacers that keep cellulose crystals apart from one another. Note that Fig. 3.1b depicts an aggregate cross section with variability in crystal cross-sectional shapes, variability in crystal orientations, and some type of organization of noncrystalline hydrated polysaccharides. It should be emphasized, however, that these characteristics of aggregates are indeed poorly understood, with limited experimental evidence available at present.

Cross sections of cellulose elementary fibrils and cellulose crystals are not necessarily coincidental. Recall that elementary fibrils are defined by their biosynthesis machinery, while crystals are defined based on the periodicities of the molecular arrangements. Cellulose synthesis and crystallization may be related but they are fundamentally distinct phenomena. In living plants, cellulose chains from *multiple* rosettes may crystallize together into *single* crystals. We here call this phenomenon *in planta co-crystallization.* Furthermore, biomass processing may also promote intermixing of *multiple* cellulose crystals, which we here call *processing co-crystallization.* Both types of co-crystallization produce crystals with more numerous cellulose chains and, therefore, larger lateral dimensions. Co-crystallization is possibly thought of as one particular, stricter type of aggregation, in which inter-crystal spacers are somehow absent and crystallographic mismatch between adjacent crystals is somehow overcome. Nevertheless, it is important to recognize that mechanisms of co-crystallization are not clearly understood yet. Despite such unknowns, cell walls of cotton are the prototypical example of *in planta* co-crystallization. Mean crystal lateral dimensions in cotton are two- to threefold greater than, for instance, in normal wood (Ioelovitch 1992; Leppänen et al. 2009), corresponding to 4–9 (i.e., 2^2–3^2) times more cellulose chains in cross sections of cotton crystals. Larger crystal lateral dimensions are usually correlated with cellulose purity, which is consistent with noncellulosic components inhibiting co-crystallization by spacing cellulose crystals (or elementary fibrils) apart from one another.

During construction of cell walls, new cellulose elementary fibrils are deposited in contact with the internal side of the already formed wall, thus increasing wall thickness in a layer-by-layer basis. A relatively thin *primary wall* is found in growing cells, while much thicker and stronger *secondary wall* is added after ceasing of cell growth (Cosgrove and Jarvis 2012). On a total mass basis, lignocellulose is made mainly of the more massive, thicker secondary walls. Layer-by-layer formation results in walls structured essentially as a sum of adhered *concentric lamellas* (Fahlén and Salmén 2002; Ding et al. 2012; Donaldson 2008). The thickness of one lamella has been reported to be about the same as aggregate lateral dimensions, i.e., 10–30 nm (Fahlén and Salmén 2002, 2005; Ding et al. 2012), reflecting that lamellas are formed by crystal aggregates. However, aggregates are essentially filaments (i.e., one-dimensional), while lamellas are sheets (i.e., two-dimensional). Secondary cell walls a few micrometers thick are formed by about hundred concentric lamellas adhered to one another. An additional, specialized type of lamella, called the *middle lamella*, is found at interfaces between cell walls of adjacent plant cells. It is noteworthy that the middle lamella is made of pectins and can be highly lignified, not being constructed in the same way as nanometric concentric lamellas of secondary cell walls. As an additional clarification, the concept of wall nanometric lamella should not be confused with the wall layers (e.g., S1, S2, and S3 layers in secondary walls of wood) (Donaldson 2008; Daniel 2009) that are usually defined based on cellulose fibril orientation.

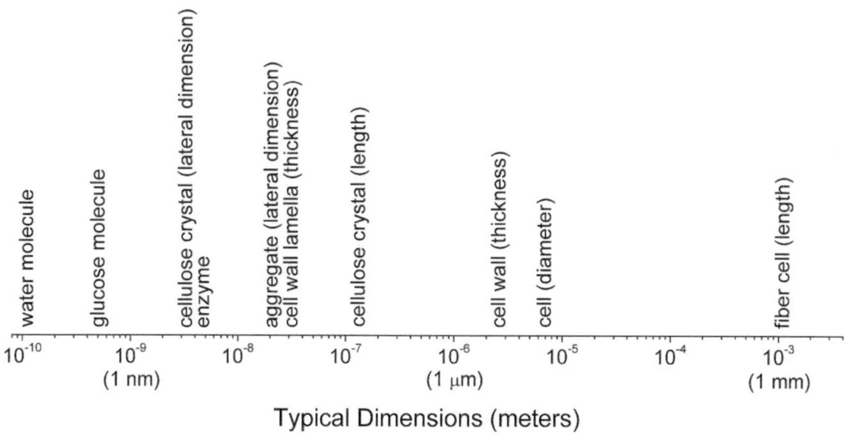

Fig. 3.2 Range of typical dimensions relevant for lignocellulose conversion. With the exception of invariant molecules, the dimensions represent ranges of variability (either natural or introduced by processing)

In summary, cellulose *crystals*, crystal *aggregates*, and wall *lamellas* are the levels of lignocellulose nanostructural hierarchy. Alternative definitions of the hierarchy are possible, such as considering elementary fibrils at the first level of hierarchy instead of cellulose crystals. Nevertheless, for an understanding of lignocellulose deconstruction the *crystal-aggregate-lamella* hierarchy is preferred because reactivity is influenced more by the crystalline and less by the fibrillar character of cellulose. The typical dimensions of lignocellulose structure are summarized in Fig. 3.2, presenting where the elements of the hierarchy fall, within and beyond the nanometric scale.

3.5 Non-crystalline Orders

In addition to crystalline cellulose, other fractions of lignocellulose—including hemicelluloses, lignin, and part of cellulose—contribute to lignocellulose nanostructure but are not organized in structures recognizable as crystals. In other words, these fractions are *non-crystalline*. Describing them as *amorphous* is perhaps inaccurate and misleading, because these fractions still present some degree of order. In secondary cell walls, which make the major fraction of lignocelluloses, all macromolecular components have preferred spatial orientation, with a degree of orientation ranked as follows: cellulose > hemicelluloses > lignin (Salmén 2014). The orientation of hemicelluloses results from the high degree of association in parallel with cellulose fibrils, while lignin orientation results from lignin deposition constrained by the polysaccharide network formed in earlier stages of cell wall development. A different situation occurs with lignin in the middle lamellae, which is formed without such spatial constraints. Therefore, middle lamella lignin is more spatially disordered and perhaps accurately qualified as amorphous (Salmén 2014).

Part of cellulose is non-crystalline, but such assignment depends on the experimental techniques used to discriminate crystalline from non-crystalline. In samples with cellulose I crystal polymorph, there are at least four distinct types of cellulose environments (enumerated below) that, depending on criteria, can be counted as non-crystalline.

1. *Cellulose chains at crystal surfaces*: Cellulose molecules have exocyclic hydroxymethyl groups (group of carbon C6). Each hydroxymethyl group can be in gauche-trans (*gt*), gauche-gauche (*gg*), or trans-gauche (*tg*) conformation. The cellulose molecules at crystal surfaces have hydroxymethyl groups in *gt* or *gg* conformations, while *tg* is the conformation in crystal cores (Atalla and VanderHart 1999; Viëtor et al. 2002). Due to the small lateral dimensions of crystals, such surface chains may correspond to large fractions (up to ≈50%) of the total cellulose. Such surface chains are represented by the outermost rectangles of each crystal in Fig. 3.1a, b.
2. *Cellulose chains in crystal-crystal interface regions*: These molecules are represented by some of the gray circles of Fig 3.1b. Such interface regions are inherent to the formation of crystal aggregates (Wickholm et al. 1998), but the molecular organization in these regions is poorly understood. Imperfect packing of the aggregated crystals and mismatch of crystallographic orientations may result in interface cellulose chains influenced by multiple and contradictory crystal orders (Driemeier and Bragatto 2013). Again, because crystals are laterally small, interface chains may correspond to an appreciable fraction of the total cellulose.
3. *Disordered domains along cellulose fibrils*: These domains have been considered critical for cellulose reactivity because they are more accessible to hydrolysis. Isolation of cellulose nanowhiskers and observation of cellulose leveling-off degree of polymerization after acid hydrolysis are evidences of this type of disorder (Nishiyama et al. 2003b). Investigation of ramie cellulose, however, indicated only 4–5 disordered residues (2.0–2.5 nm) out of every 300 glucose residues (150 nm). These are the proportions represented in Fig. 3.1c. That is, disordered domains along the fibrils were found to make only ≈1.5% of cellulose fibril length (Nishiyama et al. 2003b). Hence, these domains may be critical weak spots where cellulose deconstruction initiates, but they typically make a tiny fraction of the total cellulose mass.
4. *Independent cellulose disorder*: Cellulose can be completely decrystallized, for instance, by intense ball milling (Fink et al. 1987; Kocherbitov et al. 2008). The resulting disorder is not associated with nearby crystals of cellulose I and it is thus conceptually different from the aforementioned disorders of types 1–3, which are all somehow *dependent* on adjacent crystals of cellulose I. Therefore, this fourth type of disorder is here classified as *independent*.

Measurements of the degree of crystallinity are often employed to determine the mass fraction of crystalline cellulose in one specimen, with contents of non-crystalline matter inferred by complementarity. Although the existence of non-crystalline cellulose has been repeatedly reported based on NMR (Wickholm et al. 1998; Newman and Hemmingson 1995) and XRD (Driemeier and Bragatto 2013), interpretation of the results is often elusive. One common misconception is to attribute any non-crystalline signal to environments of the aforementioned types (3) and (4) without considering the usually dominant types (1) and (2). Another common mistake is to consider XRD peak-high Segal crystallinity index (Segal et al. 1959) as a mass fraction measurement. The Segal index has no quantitative foundation and is strongly sensitive to the lateral dimensions of cellulose I crystals (French and Cintrón 2013). Importantly, accurate measurement of cellulose degree of crystallinity employing XRD requires many considerations, including several important sources of measurement uncertainty (Driemeier and Calligaris 2011).

3.6 Nanostructural Changes Due to Thermochemical Treatments

Thermochemical treatments discussed in this section refer to traditional chemical pulping (kraft, sulfite, etc.) as well as to the so-called pretreatments developed with the aim of reducing lignocellulose resistance (*recalcitrance*) to enzymatic deconstruction. Liquid hot water, steam explosion, dilute acid, organosolv, wet oxidation, lime, and ammonia explosion are examples of leading pretreatments (Wyman et al. 2005; Alvira et al. 2010; Hendriks and Zeeman 2009). From a compositional perspective, the thermochemical treatments break, solubilize, and preferentially remove pectins, lignin, and hemicelluloses from the biomass, exposing cellulose that thus becomes much more amenable to enzymatic attack. Pulping and most leading pretreatments preserve the native polymorph cellulose I (Ioelovitch 1992; Leppänen et al. 2009; Driemeier et al. 2011; Sun et al. 2014). A notable exception is the pretreatment in ammonia, which in appropriate conditions can generate cellulose III_I (Chundawat et al. 2011; Sousa et al. 2016).

In many cases, the most important nanostructural change promoted by pretreatments is an increase in cell wall nanoscale porosity, with typical pore sizes in the 1–400 nm range. At least *two mechanisms* operate for such increase in wall porosity. *The first mechanism* is associated primarily with removal and segregation of lignin (Driemeier et al. 2016). During cell wall formation, lignin deposition in secondary walls takes place in the nanoscale pores of the previously deposited polysaccharide network, so that lignin acts as a filler of nanometric wall pores (Donaldson 2001). Lignin removal, therefore, reverses pore filling, restoring some porosity of walls. *The second mechanism* arises from wall mechanical weakening due to fragmentation of molecules that help to maintain wall structural integrity. Along with this mechanism, there are gains in nanoscale porosity mainly due to wall delamination, i.e., pore opening by separation of wall lamellas (Fahlén and Salmén 2005; Stone and Scallan 1968; Ciesielski et al. 2014). Porosity gains enabled by wall weakening may be enhanced by combining chemical treatments with mechanical processes

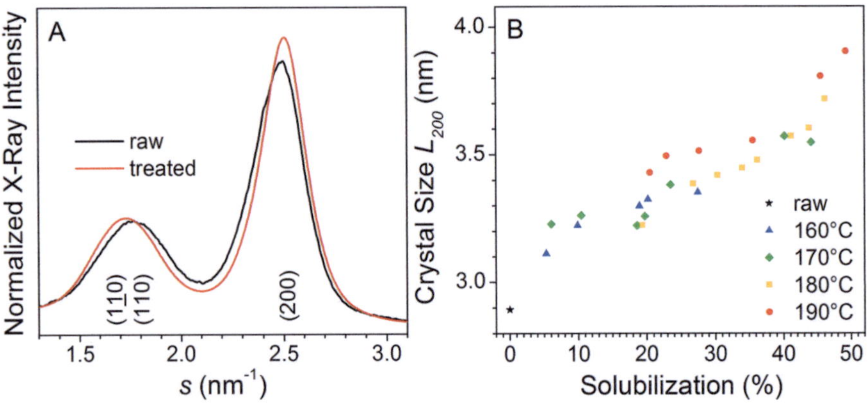

Fig. 3.3 Cellulose crystal modifications occurring in hydrothermal treatment of sugarcane bagasse. (**a**) Normalized X-ray diffraction pattern in the region of diffraction peaks $(1\underline{1}0)$, (110), and (200) (indexed as cellulose Iβ). These diffraction peaks become sharper after hydrothermal treatments, which is attributed to increasing mean crystal lateral dimensions due to cellulose co-crystallization. The horizontal axis is the magnitude of the scattering vector $s = 2\ \sin(\theta)/\lambda$. (**b**) Cellulose crystal lateral dimension, L_{200}, inferred from analysis of X-ray diffraction data and presented as function of percentage of biomass solubilized in hydrothermal treatments (performed at 160, 170, 180, and 190 °C). Data from Driemeier et al. (2015)

such as steam explosion, grinding, and refining (Ciesielski et al. 2014; Maloney and Paulapuro 1999; Wang et al. 2014; Park et al. 2016). Importantly, evaporative drying causes the collapse of lignocellulose nanoscale pores due to the cohesive action of water surface tension. Such collapse may be partly irreversible, a phenomenon known as hornification (Newman 2004; Fernandes Diniz et al. 2004).

A pivotal characteristic of most thermochemical treatments is the temperatures (120–220 °C) at which they are performed. Such temperatures are high compared to biological temperatures, allowing kinetic barriers to be overcome and dormant thermodynamic trends to manifest. Phase separation seems to be one such phenomenon, with a tendency to segregate lignin-rich and cellulose-rich fractions (Langan et al. 2014). If not solubilized, lignin-rich fractions may aggregate into nanometric globules (Langan et al. 2014), which may eventually grow into larger droplets (Donohoe et al. 2008). Dehydration of crystal aggregates appears to be another fundamental high-temperature phenomenon. In wet conditions, water penetrates within aggregates, hydrating crystal-crystal interface regions as illustrated in Fig. 3.1b (Fernandes et al. 2011; Driemeier and Bragatto 2013). This inter-crystal water is expelled as temperature is increased (Langan et al. 2014; Silveira et al. 2016), an entropy-driven process that may promote co-crystallization and further aggregation, consistent with observed increases in lateral dimensions of cellulose crystals (Ioelovitch 1992; Leppänen et al. 2009; Driemeier et al. 2011, 2015; Nishiyama et al. 2014) and in lateral dimensions of aggregates (Fahlén and Salmén 2005; Hult et al. 2001; Nishiyama et al. 2014; Foston and Ragauskas 2010), respectively. Figure 3.3 shows XRD results from sugarcane bagasse submitted to hydrothermal treatments

(Driemeier et al. 2015). XRD patterns show that treatments promote sharpening of diffraction peaks associated with crystal's lateral dimensions (Fig. 3.3a). This experimental observation is interpreted as increasing the mean lateral dimensions of cellulose crystals, which occurs progressively as treatment severity is increased (Fig. 3.3b).

It is noted that pore opening by wall delamination is the opposite of co-crystallization and aggregation. This counteraction is explained because delamination is a *separation* of adjacent lamellas, while co-crystallization and aggregation are *cohesive* phenomena taking place within each lamella.

3.7 How Nanostructure Impacts Enzymatic Digestibility

Chemical reactions for lignocellulose deconstruction depend on transport of reactants, catalysts, and products inside the porous biomass structure. Otherwise, reactions are limited to external particle surfaces. Enzymatic catalysis, in particular, depends on transport of enzymes with dimensions of ~5 nm (Fig. 3.2). Due to their sizes, enzymes cannot penetrate into cellulose crystal aggregates and enzyme action is limited to aggregates with *accessible* surfaces. Increasing the porosity of cell walls is therefore critical to enhance accessibility and, consequently, enzymatic digestibility (Grethlein 1985; Arantes and Saddler 2011; Leu and Zhu 2013; Hinkle et al. 2015). Increasing wall porosity is arguably the main benefit that physical-chemical pretreatments bring to subsequent enzymatic deconstruction.

The presence of lignin also impairs enzymatic digestion of cell wall polysaccharides. Lignin can (1) physically block enzyme access to polysaccharides and (2) adsorb enzymes unproductively (Vermaas et al. 2015). Enzyme adsorption on lignin is thought to be primarily mediated by hydrophobic interactions, but shape complementarity between enzymes and lignin fractal surfaces may also have a role (Langan et al. 2014; Petridis et al. 2011). Modification of lignin surfaces with surfactants (Eriksson et al. 2002) or pH-induced electrical charges (Lou et al. 2013) has been shown to enhance enzymatic digestibility due to lower unproductive enzyme adsorption on lignin. The presence of pectins and hemicelluloses can also block cellulase access to cellulose. Nevertheless, enzymes with proper specificity can act on pectins and hemicelluloses, breaking and removing them and thus overcoming their blocking effect and enhancing enzymatic hydrolysis of cellulose (Kumar and Wyman 2009; Hu et al. 2011; Delabona et al. 2013; Lima et al. 2016).

Most studies on enzymatic hydrolysis of cellulose are concerned with the cellulose I polymorph. Nevertheless, changes in polymorph were shown to be beneficial for hydrolysis in case of formation of cellulose hydrates (Kobayashi et al. 2012) and cellulose III_I (Chundawat et al. 2011; Sousa et al. 2016), the latter obtained through ammonia pretreatments. In spite of potential benefits, changes of cellulose polymorph occur in narrow windows of treatment conditions, thus severely constraining process optimization, a concern especially for process costs. Furthermore, many studies investigated the role that cellulose "crystallinity" could have on enzymatic hydrolysis, with results so far being contradictory or inconclusive. Types of cellulose disorder and variable crystal lateral dimensions as those discussed in earlier sections of this chapter should be clearly discriminated in order to go beyond a vague concept of "crystallinity."

3.8 Specific Nanostructural Features of Sugarcane Stalks

Lignocelluloses from different sources are thought to share the previous *qualitative* considerations about crystalline cellulose, non-crystalline ordering, nanostructural hierarchy, response to chemical treatments, and factors affecting enzymatic digestibility. However, each specific lignocellulose also has specific characteristics, some of them captured only by quantitative analyses. This section details some nanostructural features within sugarcane stalks, thus exemplifying nanostructural variability of biological origin.

The most studied nanostructural feature of wood is the orientation of cellulose fibrils within the wood fiber cell, which has a major impact on wood biomechanics (Donaldson 2008; Keckes et al. 2003). Secondary cell walls of wood are usually divided into S1, S2, and S3 layers, each layer having a characteristic fibril orientation angle to the long axis of the fiber (Donaldson 2008; Daniel 2009). The S2 layer is much thicker and has fibrils oriented close to the fiber axis, forming with the fiber axis an angle known as the microfibril angle (MFA). On the other hand, thick-walled fibers from grasses, sugarcane included, present numerous wall layers of alternate fibril orientation, differing qualitatively from wood (Parameswaran and Liese 1976; Lanfang and Terashima 1991; Burgert and Fratzl 2009). Measurements of MFA distributions in vascular bundles of sugarcane stalks showed quite broad-angle distributions, with maxima of angular distributions found between 10° and 40° (Driemeier et al. 2012). In addition, on average, MFA in sugarcane vascular bundles depends on the radial position of the bundle in the internode (Driemeier et al. 2012).

In grasses, the lignocellulose from the stalks has shown recalcitrance dependent on which plant tissue is being degraded. Differences were observed in responses to pretreatments as well as in enzymatic degradability of raw and pretreated biomass (Jung and Casler 2006; Zeng et al. 2012; Costa et al. 2013; Brienzo et al. 2014). In general, the peripheral layer of internodes, here called rind, are more recalcitrant, while the central part of internodes are less so. Furthermore, when parenchyma is separated from vascular bundles, parenchyma is observed to be less recalcitrant. Although several attempts have been made to explain differences in recalcitrance from differences in composition, tissue-specific recalcitrance is most likely due to the tissue-specific wall nanoscale porosity shown in Fig. 3.4 (Maziero et al. 2013). The observed ranking of nanoscale porosity coincides with the ranking of reactivity/digestibility. Differences along the height of the stalk (bottom, middle, top) are most likely associated with tissue maturity, with the top having more porous walls because tissues are younger and less lignified. On the other hand, porosity differences between types of tissues (rind, vascular bundles, and parenchyma from central parts of internodes) are most likely due to differences in cell wall nanoscale architectures (Maziero et al. 2013).

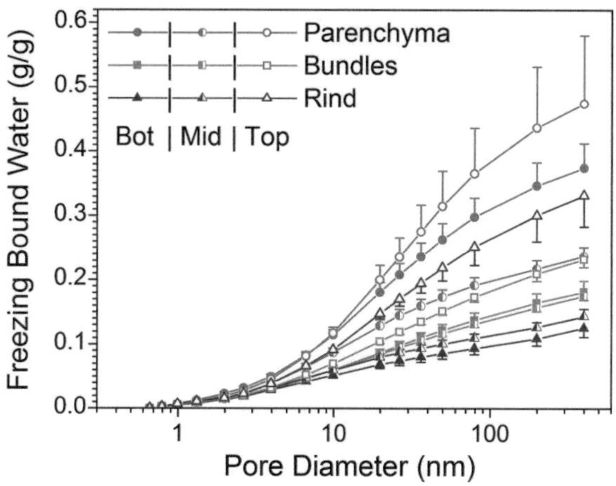

Fig. 3.4 Nanoscale porosity in water-saturated tissues dissected from sugarcane stalks. Porosity is presented as cumulative pore size distribution measured calorimetrically as freezing bound water given in units of water g per dry matter g. Peripheral ≈5 mm of internodes were called *rind*. Inner tissues of internodes were separated in vascular *bundles* and *parenchyma*. Fractions were dissected from *bottom*, *middle*, and *top* internodes from stalks of sugarcane (*see legend in graph*). Halves of symmetric error bars are omitted for better visualization. Data from Maziero et al. (2013)

3.9 Conclusions and Future Prospects

Lignocellulosic biomass is a vast renewable resource made by nature with a complex hierarchical structure that goes from molecules to, ultimately, entire plants. Nanostructural hierarchy is part of this design, going from cellulose *crystals* to cellulose crystal *aggregates* and to concentric cell wall *lamellas*. Biomass conversion into chemicals and liquid biofuels consists, essentially, in the deconstruction of this hierarchical nanostructure, releasing small molecules such as monomeric sugars that are convertible into a variety of products.

As for the future, it seems that proper measurement of lignocellulose nanoscale elements and their interrelations is still a major limitation despite important advances achieved in the last decade or so. Quantification is essential, but the characterization of nanoscale features often relies on important but insufficient qualitative insights from imaging techniques. Once quantification of a given nanostructural feature becomes possible, one should (1) drive down analytical costs to enable characterization of large sample sets, exploring biological and processing variability; (2) improve analytical precision to better perceive sample differences; and (3) apply proper analysis of multidimensional datasets to understand concretely which, how, and to what extent nanostructural features affect the desired lignocellulose conversion outcomes. If such a rationale is in place, plants and processes can be designed with consideration of nanostructure, for the purpose of harnessing the full potential of lignocellulosic biomass.

Acknowledgments The author would like to express his great gratitude to Dr. Rodrigo Silveira, Prof. Valdeir Arantes, Prof. Antônio Aprígio Curvelo, and Dr. Alfred French, who dedicated part of their valuable time to read and comment earlier versions of this chapter. The thoughtful editorial review by Dr. Amanda Souza and Prof. Marcos Buckeridge is also acknowledged. The received comments were at several points decisive for the clarity, precision, and coverage of this published chapter.

References

Akin DE (2008) Plant cell wall aromatics: influence on degradation of biomass. Biofuels Bioprod Biorefin 2:288–303. doi:10.1002/bbb

Albersheim P, Darvill A, Roberts K, Sederoff R, Staehelin A (2010) Plant cell walls: from chemistry to biology. Garland Science, Taylor and Francis Group, LLC, New York

Alvira P, Tomás-Pejó E, Ballesteros M, Negro MJ (2010) Pretreatment technologies for an efficient bioethanol production process based on enzymatic hydrolysis: a review. Bioresour Technol 101:4851–4861. doi:10.1016/j.biortech.2009.11.093

Andersson S, Serimaa R, Paakkari T et al (2003) Crystallinity of wood and the size of cellulose crystallites in Norway spruce (*Picea abies*). J Wood Sci 49:531–537. doi:10.1007/s10086-003-0518-x

Arantes V, Saddler JN (2011) Cellulose accessibility limits the effectiveness of minimum cellulase loading on the efficient hydrolysis of pretreated lignocellulosic substrates. Biotechnol Biofuels 4:3. doi:10.1186/1754-6834-4-3

Atalla RH, VanderHart DL (1984) Native cellulose: a composite of two distinct crystalline forms. Science 223:283–285

Atalla RH, VanderHart DL (1999) The role of solid state 13C NMR spectroscopy in studies of the nature of native celluloses. Solid State Nucl Magn Reson 15:1–19

Beckham GT, Matthews JF, Peters B et al (2011) Molecular-level origins of biomass recalcitrance: decrystallization free energies for four common cellulose polymorphs. J Phys Chem B 115: 4118–4127. doi:10.1021/jp1106394

Bergenstråhle M, Wohlert J, Larsson PT et al (2008) Dynamics of cellulose-water interfaces: NMR spin-lattice relaxation times calculated from atomistic computer simulations. J Phys Chem B 112:2590–2595

Brienzo M, Ferreira S, Vicentim MP et al (2014) Comparison study on the biomass recalcitrance of different tissue fractions of sugarcane culm. Bioenergy Res 7:1454–1465. doi:10.1007/s12155-014-9487-8

Buckeridge MS, dos Santos WD, Tiné MAS, De Souza AP (2016) The cell wall architecture of sugarcane and its implications to cell wall recalcitrance. In: Lam E, Carrer H, Silva JA, Kole C (eds) Compendium of bioenergy plants—sugarcane. CRC Press—Taylor & Francis Group, Boca Raton, FL, pp 31–50

Burgert I, Fratzl P (2009) Plants control the properties and actuation of their organs through the orientation of cellulose fibrils in their cell walls. Integr Comp Biol 49:69–79. doi:10.1093/icb/icp026

Chundawat SPS, Bellesia G, Uppugundla N et al (2011) Restructuring the Crystalline Cellulose Hydrogen Bond Network Enhances Its Depolymerization Rate. J Am Chem Soc 133:11163–11174. doi: 10.1021/ja2011115

Ciesielski PN, Wang W, Chen X et al (2014) Effect of mechanical disruption on the effectiveness of three reactors used for dilute acid pretreatment of corn stover Part 2: morphological and structural substrate analysis. Biotechnol Biofuels 7:47. doi: 10.1186/1754-6834-7-47

Cosgrove DJ (2005) Growth of the plant cell wall. Nat Rev Mol Cell Biol 6:850–861. doi:10.1038/nrm1746

Cosgrove DJ, Jarvis MC (2012) Comparative structure and biomechanics of plant primary and secondary cell walls. Front Plant Sci 3:1–6. doi:10.3389/fpls.2012.00204
Costa THF, Masarin F, TO B et al (2013) The enzymatic recalcitrance of inter-nodes of sugar cane hybrids with contrasting lignin contents. Ind Crop Prod 51:202–211. doi:10.1016/j.indcrop.2013.08.078
Daniel G (2009) Wood and fibre morphology. In: Ek M, Gellerstedt G, Henriksson G (eds) Wood chemistry and wood biotechnology (vol. 1). Walter de Gruyter, Berlin, pp 45–70
De Souza AP, Leite DCC, Pattathil S et al (2013) Composition and structure of sugarcane cell wall polysaccharides: implications for second-generation bioethanol production. Bioenergy Res 6:564–579. doi:10.1007/s12155-012-9268-1
Debzi EM, Chanzy H, Sugiyama J et al (1991) The Iα→Iβ transformation of highly crystalline cellulose by annealing in various mediums. Macromolecules 24:6816–6822
Delabona PS, Cota J, Hoffmam ZB et al (2013) Understanding the cellulolytic system of Trichoderma harzianum P49P11 and enhancing saccharification of pretreated sugarcane bagasse by supplementation with pectinase and α-l-arabinofuranosidase. Bioresour Technol 131:500–507. doi:10.1016/j.biortech.2012.12.105
Ding S-Y, Himmel ME (2006) The maize primary cell wall microfibril: a new model derived from direct visualization. J Agric Food Chem 54:597–606. doi:10.1021/jf051851z
Ding S-Y, Liu Y-S, Zeng Y et al (2012) How does plant cell wall nanoscale architecture correlate with enzymatic digestibility? Science 338:1055–1060
Ding S-Y, Zhao S, Zeng Y (2014) Size, shape, and arrangement of native cellulose fibrils in maize cell walls. Cellulose 21:863–871. doi:10.1007/s10570-013-0147-5
Donaldson LA (2001) Lignification and lignin topochemistry - an ultrastructural view. Phytochemistry 57:859–873.
Donaldson L (2007) Cellulose microfibril aggregates and their size variation with cell wall type. Wood Sci Technol 41:443–460
Donaldson L (2008) Microfibril angle: measurement, variation and relationships—a review. IAWA J 29:345–386
Donohoe BS, Decker SR, Tucker MP et al (2008) Visualizing lignin coalescence and mi-gration through maize cell walls following thermochemical pretreatment. Biotechnol Bioeng 101:913–925. doi:10.1002/bit.21959
Driemeier C, Bragatto J (2013) Crystallite width determines monolayer hydration across a wide spectrum of celluloses isolated from plants. J Phys Chem B 117:415–421
Driemeier C, Calligaris GA (2011) Theoretical and experimental developments for accurate determination of crystallinity of cellulose I materials. J Appl Cryst 44:184–192. doi:10.1107/S0021889810043955
Driemeier C, Francisco LH (2014) X-ray diffraction from faulted cellulose I constructed with mixed Iα–Iβ stacking. Cellulose 21(5):3161–3169. doi:10.1007/s10570-014-0390-4
Driemeier C, Pimenta MTB, Rocha GJM et al (2011) Evolution of cellulose crystals during prehydrolysis and soda delignification of sugarcane lignocellulose. Cellulose 18:1509 1519. doi:10.1007/s10570-011-9592-1
Driemeier C, Santos WD, Buckeridge MS (2012) Cellulose crystals in fibrovascular bundles of sugarcane culms: orientation, size, distortion, and variability. Cellulose 19:1507–1515. doi:10.1007/s10570-012-9743-z
Driemeier C, Mendes FM, Santucci BS, Pimenta MTB (2015) Cellulose co-crystallization and related phenomena occurring in hydrothermal treatment of sugarcane bagasse. Cellulose 22(4):2183–2195. doi:10.1007/s10570-015-0638-7
Driemeier C, Oliveira MM, Curvelo AAS (2016) Lignin contributions to the nanoscale porosity of raw and treated lignocelluloses as observed by calorimetric thermoporometry. Ind Crop Prod 82:114–117. doi:10.1016/j.indcrop.2015.11.084
Eriksson T, Börjesson J, Tjerneld F (2002) Mechanism of surfactant effect in enzymatic hydrolysis of lignocellulose. Enzyme Microb Technol 31:353–364. doi:10.1016/S0141-0229(02)00134-5
Fahlén J, Salmén L (2002) On the lamellar structure of the tracheid cell wall. Plant Biol 4:339–345. doi:10.1055/s-2002-32341

Fahlén J, Salmén L (2005) Pore and matrix distribution in the fiber wall revealed by atomic force microscopy and image analysis. Biomacromolecules 6:433–438
Fernandes Diniz JMB, Gil MH, Castro JAAM (2004) Hornification—its origin and interpretation in wood pulps. Wood Sci Technol 37:489–494. doi:10.1007/s00226-003-0216-2
Fernandes AN, Thomas LH, Altaner CM et al (2011) Nanostructure of cellulose microfibrils in spruce wood. Proc Natl Acad Sci 108:E1195–E1203. doi:10.1073/pnas.1108942108
Fink H-P, Philipp B, Paul D et al (1987) The structure of amorphous cellulose as revealed by wide-angle X-ray scattering. Polymer (Guildf) 28:1265–1270. doi:10.1016/0032-3861(87)90435-6
Foston M, Ragauskas AJ (2010) Changes in lignocellulosic supramolecular and ultrastructure during dilute acid pretreatment of Populus and switchgrass. Biomass Bioenergy 34:1885–1895. doi:10.1016/j.biombioe.2010.07.023
Fratzl P, Weinkamer R (2007) Nature's hierarchical materials. Prog Mater Sci 52:1263–1334. doi:10.1016/j.pmatsci.2007.06.001
French AD (2014) Idealized powder diffraction patterns for cellulose polymorphs. Cellulose 21: 885–896. doi:10.1007/s10570-013-0030-4
French AD, Cintrón MS (2013) Cellulose polymorphy, crystallite size, and the Segal crystallinity index. Cellulose 20:583–588. doi:10.1007/s10570-012-9833-y
Grethlein HE (1985) The effect of pore size distribution on the rate of enzymatic hydrolysis of cellulosic substrates. Nat Biotechnol 3:155–160
Hendriks ATWM, Zeeman G (2009) Pretreatments to enhance the digestibility of lignocellulosic biomass. Bioresour Technol 100:10–18. doi:10.1016/j.biortech.2008.05.027
Henriksson G (2009) Lignin. In: Ek M, Gellerstedt G, Henriksson G (eds) Wood chemistry and wood biotechnology (vol. 1). Walter de Gruyter, Berlin, pp 121–145
Hinkle JD, Ciesielski PN, Gruchalla K et al (2015) Biomass accessibility analysis using electron tomography. Biotechnol Biofuels 8:212. doi:10.1186/s13068-015-0395-8
Hori R, Wada M (2005) The thermal expansion of wood cellulose crystals. Cellulose 12:479–484. doi:10.1007/s10570-005-5967-5
Horii F, Yamamoto H, Kitamaru R et al (1987) Transformation of native cellulose crystals induced by saturated steam at high temperatures. Macromolecules 20:2946–2949
Hu J, Arantes V, Saddler JN (2011) The enhancement of enzymatic hydrolysis of lignocellulosic substrates by the addition of accessory enzymes such as xylanase: is it an additive or synergistic effect? Biotechnol Biofuels 4:36. doi:10.1186/1754-6834-4-36
Hult E-L, Larsson PT, Iversen T (2001) Cellulose fibril aggregation—an inherent property of kraft pulps. Polymer 42:3309–3314
Ioelovitch M (1992) Zur übermolekularen Struktur von nativen und isolierten Cellulosen. Acta Polym 43:110–113
Jarvis M (2003) Cellulose stacks up. Nature 426:611–612
Jung HG, Casler MD (2006) Maize stem tissues: impact of development on cell wall degradability. Crop Sci 46:1801–1809. doi:10.2135/cropsci2006.02-0086
Keckes J, Burgert I, Frühmann K et al (2003) Cell-wall recovery after irreversible deformation of wood. Nat Mater 2:810–814. doi:10.1038/nmat1019
Klemm D, Heublein B, Fink H-P, Bohn A (2005) Cellulose: fascinating biopolymer and sustainable raw material. Angew Chem Int Ed 44:3358–3393. doi:10.1002/anie.200460587
Kobayashi K, Kimura S, Kim U-J et al (2012) Enzymatic hydrolysis of cellulose hydrates. Cellulose 19:967–974. doi:10.1007/s10570-012-9696-2
Kocherbitov V, Ulvenlund S, Kober M et al (2008) Hydration of microcrystalline cellulose and milled cellulose studied by sorption calorimetry. J Phys Chem B 112:3728–3734
Kumar R, Wyman CE (2009) Effect of xylanase supplementation of cellulase on digestion of corn stover solids prepared by leading pretreatment technologies. Bioresour Technol 100:4203–4213. doi:10.1016/j.biortech.2008.11.057
Lanfang H, Terashima N (1991) Formation and structure of lignin in monocotyledons. IV. Deposition process and structural diversity of the lignin in the cell wall of sugarcane and rice plant studied by ultraviolet microscopic spectroscopy. Holzforschung 45:191–198

Langan P, Nishiyama Y, Chanzy H (2001) X-ray structure of mercerized cellulose II at 1 Å resolution. Biomacromolecules 2:410–416

Langan P, Petridis L, O'Neill HM et al (2014) Common processes drive the thermochemical pretreatment of lignocellulosic biomass. Green Chem 16:63. doi:10.1039/c3gc41962b

Lehtiö J, Sugiyama J, Gustavsson M et al (2003) The binding specificity and affinity determinants of family 1 and family 3 cellulose binding modules. Proc Natl Acad Sci U S A 100:484–489. doi:10.1073/pnas.212651999

Leppänen K, Andersson S, Torkkeli M et al (2009) Structure of cellulose and microcrystal-line cellulose from various wood species, cotton and flax studied by X-ray scattering. Cellulose 16: 999–1015. doi:10.1007/s10570-009-9298-9

Leu S-Y, Zhu JY (2013) Substrate-related factors affecting enzymatic saccharification of lignocelluloses: our recent understanding. Bioenergy Res 6:405–415. doi:10.1007/s12155-012-9276-1

Lima EA, Machado CB, Zanphorlin LM et al (2016) GH53 endo-beta-1,4-galactanase from a newly isolated Bacillus licheniformis CBMAI 1609 as an enzymatic cocktail supplement for biomass saccharification. Appl Biochem Biotechnol. doi:10.1007/s12010-016-2003-1

Lou H, Zhu JY, Lan TQ et al (2013) pH-Induced lignin surface modification to reduce non-specific cellulase binding and enhance enzymatic saccharification of lignocelluloses. ChemSusChem 6:919–927. doi:10.1002/cssc.201200859

Maloney TC, Paulapuro H (1999) The formation of pores in the cell wall. J Pulp Pap Sci 25: 430–436

Mazeau K (2011) On the external morphology of native cellulose microfibrils. Carbohydr Polym 84:524–532

Maziero P, Jong J, Mendes FM et al (2013) Tissue-specific cell wall hydration in sugarcane stalks. J Agric Food Chem 61:5841–5847. doi:10.1021/jf401243c

Newman RH (2004) Carbon-13 NMR evidence for cocrystallization of cellulose as a mechanism for hornification of bleached kraft pulp. Cellulose 11:45–52. doi:10.1023/B:CELL.0000014768.28924.0c

Newman RH, Hemmingson JA (1995) Carbon-13 NMR distinction between categories of molecular order and disorder in cellulose. Cellulose 2:95–110

Newman RH, Hill SJ, Harris PJ (2013) Wide-angle x-ray scattering and solid-state nuclear magnetic resonance data combined to test models for cellulose microfibrils in mung bean cell walls. Plant Physiol 163:1558–1567. doi:10.1104/pp.113.228262

Nishiyama Y (2009) Structure and properties of the cellulose microfibril. J Wood Sci 55:241–249. doi:10.1007/s10086-009-1029-1

Nishiyama Y, Langan P, Chanzy H (2002) Crystal structure and hydrogen bonding system in cellulose Iβ from synchrotron X-ray and neutron fiber diffraction. J Am Chem Soc 124:9074–9082. doi:10.1021/ja037055w

Nishiyama Y, Sugiyama J, Chanzy H, Langan P (2003a) Crystal structure and hydrogen bonding system in cellulose Iα from synchrotron X-ray and neutron fiber diffraction. J Am Chem Soc 125:14300–14306. doi:10.1021/ja037055w

Nishiyama Y, Kim U-J, Kim D-Y et al (2003b) Periodic disorder along ramie cellulose microfibrils. Biomacromolecules 4:1013–1017

Nishiyama Y, Langan P, O'Neill H et al (2014) Structural coarsening of aspen wood by hydrothermal pretreatment monitored by small- and wide-angle scattering of X-rays and neutrons on oriented specimens. Cellulose 21:1015–1024. doi:10.1007/s10570-013-0069-2

O'Sullivan AC (1997) Cellulose: the structure slowly unravels. Cellulose 4:173–207

Parameswaran N, Liese W (1976) On the fine structure of bamboo fibres. Wood Sci Technol 10: 231–246

Park J, Jones B, Koo B et al (2016) Use of mechanical refining to improve the production of low-cost sugars from lignocellulosic biomass. Bioresour Technol 199:59–67. doi:10.1016/j.biortech.2015.08.059

Petridis L, Pingali SV, Urban V et al (2011) Self-similar multiscale structure of lignin revealed by neutron scattering and molecular dynamics simulation. Phys Rev E 83:061911. doi:10.1103/PhysRevE.83.061911

Salmén L (2014) Wood morphology and properties from molecular perspectives. Ann Forest Sci. doi:10.1007/s13595-014-0403-3
Scheller HV, Ulvskov P (2010) Hemicelluloses. Annu Rev Plant Biol 61:263–289. doi:10.1146/annurev-arplant-042809-112315
Segal L, Creely JJ, Martin AE Jr, Conrad CM (1959) An empirical method for estimating the degree of crystallinity of native cellulose using the X-ray diffractometer. Textil Res J 29:786–794
Silveira RL, Stoyanov SR, Gusarov S et al (2013) Plant biomass recalcitrance: effect of hemicellulose composition on nanoscale forces that control cell wall strength. J Am Chem Soc 135(51):19048–19051. doi:10.1021/ja405634k
Silveira RL, Stoyanov SR, Kovalenko A, Skaf MS (2016) Cellulose Aggregation under Hydrothermal Pretreatment Conditions. Biomacromolecules acs.biomac.6b00603. doi:10.1021/acs.biomac.6b00603
Somerville C (2006) Cellulose synthesis in higher plants. Annu Rev Cell Dev Biol 22:53–78. doi:10.1146/annurev.cellbio.22.022206.160206
Sousa LC, Jin M, Chundawat SPS et al (2016) Next-generation ammonia pretreatment enhances cellulosic biofuel production. Energ Environ Sci. doi:10.1039/C5EE03051J
Stone JE, Scallan AM (1968) A structural model for the cell wall of water-swollen wood pulp fibres based on their accessibility to macromolecules. Cellulose Chem Technol 2:343–358
Sun Q, Foston M, Sawada D et al (2014) Comparison of changes in cellulose ultrastructure during different pretreatments of poplar. Cellulose 21:2419–2431. doi:10.1007/s10570-014-0303-6
Vermaas J V., Petridis L, Qi X et al (2015) Mechanism of lignin inhibition of enzymatic biomass deconstruction. Biotechnol Biofuels :1–16. doi: 10.1186/s13068-015-0379-8
Viëtor RJ, Newman RH, Ha M-A et al (2002) Conformational features of crystal-surface cellulose from higher plants. Plant J 30:721–731
Wada M (2002) Lateral thermal expansion of cellulose Iβ and III_I polymorphs. J Polym Sci B 40:1095–1102. doi:10.1002/polb.10166
Wada M, Chanzy H, Nishiyama Y, Langan P (2004a) Cellulose III_I crystal structure and hydrogen bonding by synchrotron X-ray and neutron fiber diffraction. Macromolecules 37:8548–8555. doi:10.1021/ma0485585
Wada M, Heux L, Sugiyama J (2004b) Polymorphism of cellulose I family: reinvestigation of cellulose IV_I. Biomacromolecules 5:1385–1391. doi:10.1021/bm0345357
Wada M, Nishiyama Y, Langan P (2006) X-ray structure of ammonia-cellulose I: new insights into the conversion of cellulose I to cellulose IIII. Macromolecules 39:2947–2952. doi:10.1021/ma060228s
Wada M, Heux L, Nishiyama Y, Langan P (2009) X-ray crystallographic, scanning micro-probe X-ray diffraction, and cross-polarized/magic angle spinning 13C NMR studies of the structure of cellulose III_{II}. Biomacromolecules 10:302–309. doi:10.1021/bm8010227
Wang W, Chen X, Donohoe BS et al (2014) Effect of mechanical disruption on the effec-tiveness of three reactors used for dilute acid pretreatment of corn stover part 1: chemical and physical substrate analysis. Biotechnol Biofuels 7:57. doi:10.1186/1754-6834-7-57
Wickholm K, Larsson PT, Iversen T (1998) Assignment of non-crystalline forms in cellulose I by CP/MAS ^{13}C NMR spectroscopy. Carbohydr Res 312:123–129
Wyman CE, Dale BE, Elander RT et al (2005) Coordinated development of leading biomass pretreatment technologies. Bioresour Technol 96:1959–1966. doi:10.1016/j.biortech.2005.01.010
Xiao C, Anderson CT (2013) Roles of pectin in biomass yield and processing for biofuels. Front Plant Sci 4:67. doi:10.3389/fpls.2013.00067
Zeng M, Ximenes E, Ladisch MR et al (2012) Tissue-specific biomass recalcitrance in corn stover pretreated with liquid hot-water: enzymatic hydrolysis (part 1). Biotechnol Bioeng 109:390–397. doi:10.1002/bit.23337
Zhang T, Mahgsoudy-Louyeh S, Tittmann B, Cosgrove DJ (2014) Visualization of the nanoscale pattern of recently-deposited cellulose microfibrils and matrix materials in never-dried primary walls of the onion epidermis. Cellulose 21:853–862. doi:10.1007/s10570-013-9996-110.1021/bm040068x

Chapter 4
Phenolic Compounds in Plants: Implications for Bioenergy

Dyoni Matias de Oliveira, Aline Finger-Teixeira, Denis Leandro de Freitas, Gabriela Ellen Barreto, Rogério Barbosa de Lima, Anderson Ricardo Soares, Osvaldo Ferrarese-Filho, Rogério Marchiosi, and Wanderley Dantas dos Santos

Abstract Lignin is a copolymer of three main hydroxycinnamyl alcohols identified as *p*-hydroxyphenyl (H), guaiacyl (G), and syringyl (S) units. The highly condensed matrix (core lignin) may also be associated with low-molecular phenolics as hydroxycinnamic acids mainly *p*-coumaric, caffeic, ferulic, and sinapic acids dubbed noncore lignins. Lignin confers hydrophobicity and mechanical and chemical strength for tissues, providing a barrier against the attack of pathogens and herbivores. The content and composition of lignin are strongly affected by biotic and abiotic stresses. Besides core and noncore lignin, free phenolic compounds perform a relevant activity in response to plant stresses. The toxicity of allelochemicals is partially due to their ability to bind and inhibit enzyme activities. The presence of lignin imposes a physical barrier to the action of enzymes in saccharification of plant cell wall polysaccharides to the production of cellulosic ethanol. The presence of endogenous phenolic compounds as well as treatments to degrade lignin, in turn, release phenolic compounds that adsorb and inhibit cellulases, xylanases, and accessory enzymes. This chapter provides basic information on phenolic compounds of interest to support the sustainable use of alga and plant biomasses as raw materials for the production of biofuels discussing the main approaches ongoing to reduce their negative impact in biomass saccharification.

Keywords Lignin • Phenylpropanoids • Allelopathy • Plant defense • Plant stress

D.M. de Oliveira • D.L. de Freitas • G.E. Barreto • R.B. de Lima • A.R. Soares
O. Ferrarese-Filho • R. Marchiosi • W.D. dos Santos (✉)
Plant Biochemistry Laboratory, Department of Biochemistry, State University of Maringá,
Av. Colombo, Maringá, 87020-900 Paraná, Brazil
e-mail: wdsantos@uem.br

A. Finger-Teixeira
Department of Biology, Campus of Paranavai, Parana Federal Institute of Education,
Science and Technology, Paraná, Brazil

M.S. Buckeridge, A.P. De Souza (eds.), *Advances of Basic Science for Second Generation Bioethanol from Sugarcane*, DOI 10.1007/978-3-319-49826-3_4

4.1 Introduction

The use of lignocellulosic biomass as a renewable energy source can provide a significant contribution to the development of a sustainable industrial society. The lignocellulose consists mainly of polysaccharides and structural phenylpropanoids. The lignin and hydroxycinnamates such as ferulic and *p*-coumaric acids (noncore lignin) are critical components of supporting tissues, and they also perform an essential role in plant defense against plant predators. The roles of noncore lignin are particularly relevant in grasses, which are the primary sources of biomass for production of biofuels.

Just as these compounds inhibit the action of enzymes from pathogens and herbivores, they also reduce the efficiency of saccharification of lignocellulosic biomass, currently considered an essential step in the production of second-generation ethanol (E2G). Therefore, a significant effort has been devoted to reduce the synthesis of these compounds in a controlled manner, as well as efficiently remove them postharvest to improve saccharification. Thus, a basic knowledge of structural phenylpropanoids is a fundamental requirement for physiologists, biochemistries, and engineers interested in the research and development of technologies for the E2G production.

In this chapter, we attempt to meet this demand by introducing the major chemical characteristics, metabolic pathways, and biological roles of core and noncore lignin. Also, the chapter presents a brief, but clarifying review of the state-of-the-art research aimed at producing more friendly raw materials for the production of E2G.

4.2 Lignin Emergence and Composition

The plant cell wall consists mainly of cellulose, a polysaccharide formed from polymerization of β-glucopyranose molecules. Associated with cellulose, other polysaccharides occur as hemicelluloses and pectins. At the end of cell growth, some cells can deposit lignin monomers, which are polymerized in muro conferring mechanical and chemical resistance to the wall. The emergence of vascular plants, about 400 millions of years ago, allowed plant kingdom to conquer the continental environment (Weng and Chapple 2010). Lignin is responsible for the tensile strength that enabled plants to transport water from the roots to the leaves, and sustain large stems and high canopies. Lignin further provides resistance against the attack of pathogens and herbivores reducing enzyme access to the wall polysaccharides and the cell content. Therefore, lignin is of great industrial importance either to provide resistance and quality to woods and papers or to hinder the production of cellulose, forage digestibility, and production of cellulosic ethanol (dos Santos et al. 2014). Besides the production of several carbohydrate polymers, wall biosynthesis ended up integrating phenylpropanoid metabolism into the synthesis of lignin, leading to the establishment of a Glycomic Code (Buckeridge and De Souza 2014).

Angiosperms synthesize lignin by polymerizing three main monolignols: *p*-coumaryl alcohol, coniferyl alcohol, and sinapyl alcohol (Moreira-Vilar et al. 2014). By historical causes, once polymerized, the monomeric composition is identified as *p*-hydroxyphenyl (H unit), guaiacyl (G unit), and syringyl (S unit). These monomers are irregularly linked by a variety of ether linkages, the most common being the arylglycerol-β-aryl bond (β-O-4), but many other linkages are present (Fig. 4.1).

The composition of lignin varies according to the type of cell and plant. Softwoods (gymnosperms or conifers) present 27–33% lignin made mainly by G units (*c.* 90%). Hardwoods (angiosperms) have from 18 to 25% lignin, with similar amounts of units G and S and only a small amount of H units. In turn, grasses, as sugarcane bagasse, may present similar content of H, G, and S units (Moreira-Vilar et al. 2014).

Lignin is insoluble in most organic solvents, so its isolation is difficult. When there is a separation, the molecular structure is generally compromised. Although it is not possible to extract the lignin without degrading it, it is estimated that the molecular weight is in the range of 1,000–20,000 units. Fragments of isolated lignin present dark color and are easily oxidized due to its high aromatic content. They are relatively stable in acidic solutions and soluble in hot aqueous bases.

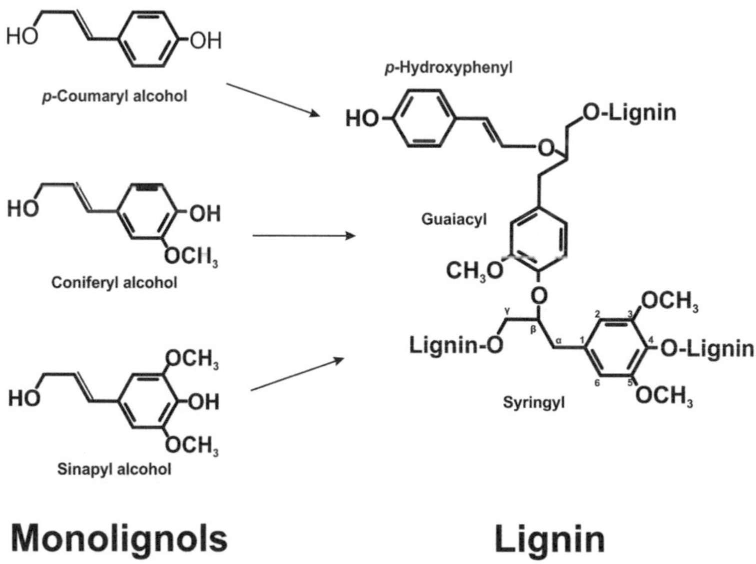

Fig. 4.1 Monolignols and lignin fragment showing the main linkage β-O-4. The origin of the monomeric units may be inferred by the substituent functions of the aromatic rings

4.3 Lignin Biosynthesis

Both core and noncore lignin components are produced in the phenylpropanoid pathway. This secondary metabolism begins with the activity of phenylalanine ammonia lyase (PAL) on phenylalanine producing cinnamic acid and ammonia. The action of the cinnamate 4-hydroxylase (C4H) on cinnamate releases *p*-coumaric acid, the first phenolic compound to be produced in the biosynthetic route. In grasses, PAL enzyme is also active on tyrosine receiving the alternative name of tyrosine ammonia lyase (TAL). It catalyzes the conversion of tyrosine directly in *p*-coumaric acid. The activation of *p*-coumaric acid with coenzyme A by 4-coumarate coenzyme A ligase (4-CL) releases the *p*-coumaroyl-CoA ester, which can follow two different routes. The action of cinnamoyl-CoA reductase (CCR) produces *p*-coumaraldehyde, which is substrate to cinnamyl alcohol dehydrogenase (CAD). The enzyme produces *p*-coumaroyl alcohol, the simplest monolignol.

The *p*-coumaroyl-CoA thioester is also substrate for hydroxycinnamate-CoA transferase (HCT), which substitutes the thioesterification with coenzyme A to an esterification with shikimic or quinic acids. The esterification with shikimic or quinic acids, produced in the primary metabolism, is thought to perform a role in the regulation of the pathway. The *p*-coumaroyl shikimate/quinate is the substrate of coumarate 3′-hydroxylase (C3′H) that catalyzes the biosynthesis of caffeoyl ester. The enzyme HCT promotes the transesterification of caffeoyl shikimate/quinate to produce caffeoyl-CoA, the precursor of caffeic acid by the action of the enzyme caffeoyl shikimate/quinate esterase (CSE) and feruloyl-CoA by the action of the enzyme caffeoyl-CoA orthomethyl transferase (CCoAOMT). Caffeic acid can generate ferulic acid by the action of caffeate orthomethyl transferase (COMT). The enzyme CCR catalyzes the reductive cleavage of feruloyl-CoA producing coniferaldehyde. The activity of coniferaldehyde dehydrogenase (CALDH) on coniferaldehyde also produces ferulic acid (Fig. 4.2, right). The enzyme CCR is also active on *p*-coumaroyl-CoA, releasing *p*-coumaraldehyde. Cinnamyl alcohol dehydrogenase then reduces the coniferaldehyde to coniferyl alcohol monolignol, the most abundant component of lignin.

The enzyme ferulate 5-hydroxylase catalyzes the hydroxylation of coniferaldehyde to give rise to 5-hydroxyconiferaldehyde, another substrate of COMT, which in this case produces sinapaldehyde. Once again, CAD reduces the sinapaldehyde to sinapyl alcohol monolignol. At least some plants present reductase specific for sinapaldehyde, therefore dubbed SAD (Fig. 4.2).

4.4 Roles of Core and Noncore Lignin in the Cell Wall Architecture

Lignocellulosic biomass is largely made up of the secondary cell walls with relevant differences in the proportion of their constituents among plant species and tissues (Carpita et al. 2001). Cellulose, the main component of the cell wall, is a

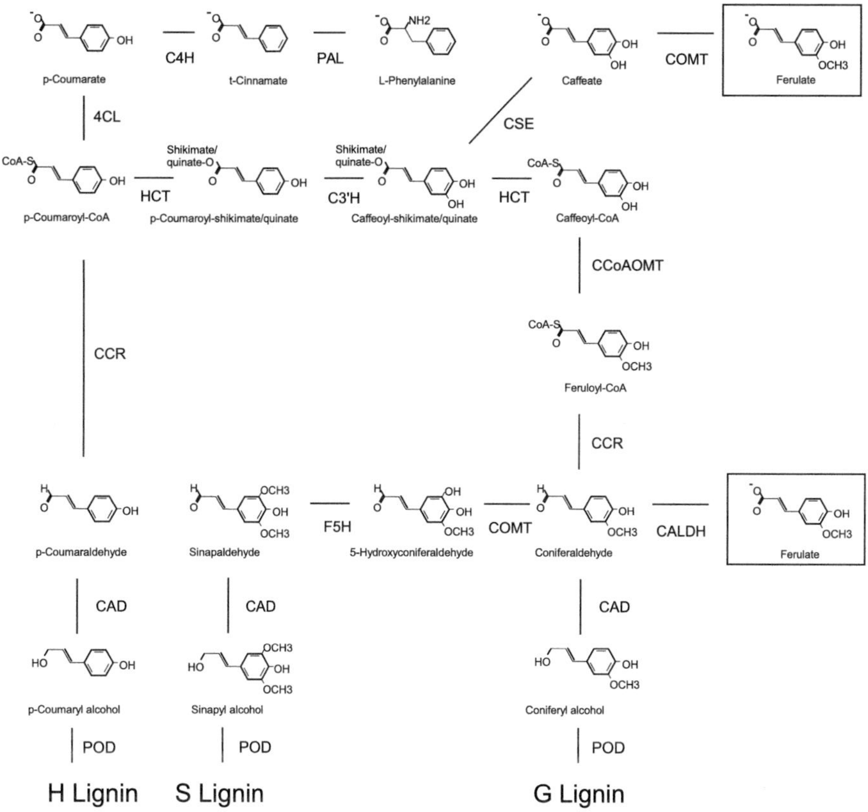

Fig. 4.2 Simplified phenylpropanoid pathway responsible for the biosynthesis of core and non-core lignin precursors (see text for detailed description of the enzymes acronyms—*arrows*)

homopolysaccharide consisting of a long and linear chain of β(1 → 4)-linked glucose units. Each cellulose molecule is tightly bound to other molecules by means of multiple hydrogen bonds producing insoluble, rigid, and crystalline microfibrils (Carpita and McCann 2000). The presence of hemicelluloses with distinct physical and chemical properties avoids celluloses microfibrils to collapse to produce macrofibrils. The richer topology of these frequently branched hemicelluloses allows them to link only occasionally with cellulose microfibrils. Intermittent pattern of free and bound regions from hemicellulose with cellulose results in the cross-linking of microfibrils (Carpita 1996). The overall architectures of cell walls consist of a network of cellulose fibers surrounded by a matrix of noncellulosic polysaccharides. Besides hemicelluloses, cell walls present pectins a complex group of heteropolysaccharides branched with acidic sugars as glucuronic and galacturonic acids, with a capacity of adsorbing high amounts of water. These polysaccharides form a gel, which is involved in cell-cell adhesion, pore sizing, pH control, and cation trapping (Buckeridge et al. 2010).

Primary cell walls can be divided into two broad categories: types I and II. Type I is found in dicots, non-commelinoid angiosperms (e.g., aroids, alismatids, and liliods) and gymnosperms. Type I cell walls consist of cellulose fibers encased in a network of xyloglucan (XyG), pectin, and structural proteins. XyG is composed of a glucose backbone branched with xylose which in turn may be substituted with galactose and fucose (Carpita and Gibeaut 1993). Type II cell walls, found only in the commelinoid monocotyledons (e.g., grasses, sedges, rushes, and gingers), are composed of cellulose fibers encased in glucuronoarabinoxylan (GAX) matrix. Those GAX may be highly branched with hydroxycinnamates, such as ferulic acid (FA), while lignin can present a high content of *p*-coumaric acids. In addition, the cell walls of grasses and some related families in the Poales order such as sugarcane contain significant quantities of mixed-linkage β-glucans (Carpita and Gibeaut 1993; Vogel 2008). The most relevant characteristic of type II cell walls is the abundance of GAX. It is composed of a core chain of xylan branched with arabinose and glucuronic acid, with the arabinose residues esterified with feruloyl residues. Type II cell walls have a low content of pectin and structural proteins (for details about sugarcane cell wall composition, see Chap. 2).

Produced in the phenylpropanoid pathway together with monolignols (Fig. 4.2), FA and ferulic esters are polymerized oxidatively by the action of peroxidases. The reaction produces dehydrodimers such as 5-5′-dehydrodiferulic acid, 8-*O*-4′-dehydrodiferulic acid, and 8-5′-dehydrodiferulic acid, along with higher oligomers. In grasses, FA might branch GAX to produce FA-GAX, a polysaccharide believed to occur only in these species and perform a set of distinguished structural and physiological roles in their cell walls (Carpita et al. 2001; dos Santos et al. 2008a). Ferulic acid occurs in higher concentrations in type II cell walls of commelinoids, e.g., maize grain (*Zea mays*) 20.66 mg g^{-1} cell wall, sugarcane (*Saccharum officinarum*) 8.0 to 17.0 3.51 mg g^{-1}, barley (*Hordeum vulgare*) 3.51 mg g^{-1}, and perennial ryegrass (*Lolium perenne*) 6.03 mg g^{-1}. However, it might also occur in high concentration in type I cell walls of "core" Caryophyllales as beetroot (*Beta vulgaris*) 6.93 mg g^{-1}. In non-commelinoid monocots and dicots, FA occurs in much lower concentrations: e.g., asparagus (*Asparagus officinalis*) 0.078–0.096 mg g^{-1} and onion (*Allium cepa*) 0.007 mg g^{-1} (Harris and Trethewey 2010).

Secondary cell walls contain lignin that is deposited inside of the primary cell walls. Secondary cell walls are prominent features of xylem, fibers, and sclerenchyma cell. This rigid structure of cell wall is a frontline barrier against microorganisms and pathogens. Beyond being hard to digest, lignin fractions adsorb hydrolytic proteins reducing the access of enzymes to the polysaccharides (Huang et al. 2011). In mature cells, lignin forms a highly hydrophobic matrix of C–C- and C–O–C-linked phenylpropanoids, mainly coniferyl, sinapyl, and *p*-coumaroyl alcohols. Respective residues in lignin are dubbed G, S, and H units. The amount of lignin, as well as its monomeric composition, is ontogenetic-tissue-species dependent. Lignin confers high hydrophobicity and mechanical resistance to cell walls required by xylem vessels to perform the capillary transport of water and for fibers support the massive habit of trees (Pedersen et al. 2005; Zobiole et al. 2010). In lignified secondary cell walls, the FA ester linked to GAX is a nucleation site for lignin polymerization through ether bounds, anchoring lignin to polysaccharide moiety (Renger and Steinhart 2000; Carpita et al. 2001).

4.5 The Role of BAHD Acyl-CoA Transferases in GAX Feruloylation

Using bioinformatics approach, Mitchell et al. (2007) compared expressed sequence tags (ESTs) from an orthologous group of grasses and dicots to identify clades where genes are differentially expressed in grasses. One of the clades showing the greatest bias in gene expression was within the BAHD acyl-CoA transferase gene family (also referred as PFAM PF02458 family). BAHD family is named after the first four characterized members: benzylalcohol *O*-acetyltransferase from *Clarkia breweri* (**B**EAT); anthocyanin *O*-hydroxycinnamoyl transferases from *Petunia*, *Senecio*, *Gentiana*, *Perilla*, and *Lavandula* (**A**HCTs); anthranilate *N*-hydroxycinnamoyl/benzoyltransferase from *Dianthus caryophyllus* (**H**CBT); and deacetylvindoline 4-*O*-acetyltransferase from *Catharanthus roseus* (**D**AT). BAHD proteins are predicted to be cytosolic based on the protein sequences, and this is the localization for all known members of the family (D'Auria 2006).

The predicted role of BAHD proteins is supported by experiments using RNAi in rice, resulting in significant decreases (−19%) in cell wall-bound FA residues (Piston et al. 2010). In another study, rice lines with one BAHD gene upregulated increased the amount of *p*-coumaroyl ester linked to GAX (Bartley et al. 2013). These studies provide strong circumstantial support for the hypothesis that BAHD genes are responsible for GAX feruloylation, although a definitive evidence is still lacking. Putative genes within a grass-specific clade of the BAHD acyl-CoA transferase superfamily have been identified as being responsible for the ester linkage of FA to GAX. The mechanism about which FA is esterified into GAX is not elucidated, and genes and enzymes responsible for feruloylation of GAX are still unknown. There are two possible potential mechanisms leading to arabinoxylan feruloylation (Fig. 4.3). In the first, GAX may be synthesized in the Golgi apparatus and then feruloylated, by feruloyl transferases, using feruloyl-CoA transported from cytosol as an FA donor (Buanafina 2009; Molinari et al. 2013). The second putative mechanism involves a BAHD acyl-CoA transferase cytosolic responsible for feruloylation of UDP-arabinosyl to form FA-Ara-UDP, which is substrate for GAX feruloylation. In this mechanism it is suggested that a glycosyltransferase, present in Golgi lumen, introduces the ferulate-arabinosyl (FA-Ara) into the nascent GAX chains with feruloyl-CoA acting as a donor substrate. The last steps in both mechanisms occur in the Golgi apparatus, and the feruloylated glucuronoarabinoxylan (FA-GAX) can be oxidatively coupled, forming dehydroferulic acid dimers. FA-GAX and their coupled derivatives are transported into cell wall by the Golgi secretion (Buanafina 2009; Anders et al. 2012; Burton and Fincher 2012).

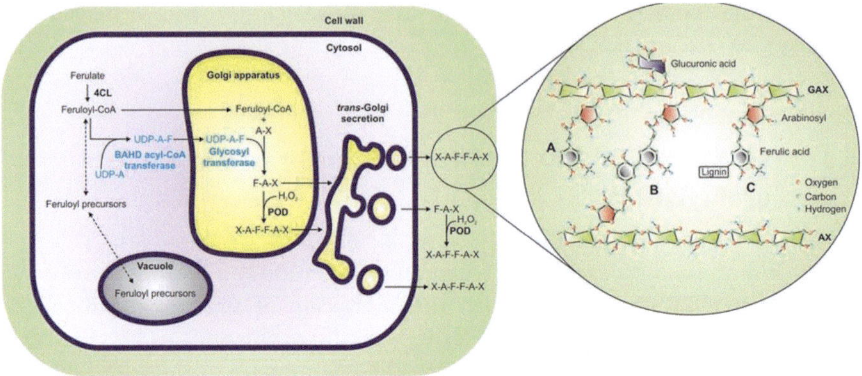

Fig. 4.3 Routes for GAX feruloylation. *Blue text* shows possible pathways for feruloylation based on cytosolic location of BAHD proteins implicated in feruloylation. *A* arabinosyl, *FA* feruloyl group, *F-A-X* feruloylated arabinoxylan, *POD* peroxidase, *Ara-UDP* UDP-arabinofuranose, *X* xylan polymer. *Right*: Schematic representation of FA-GAX and FA-GAX cross-linked by dehydrodiferulic acid (Buanafina 2009; Mastrangelo et al. 2009; Oliveira et al. 2014)

Esterification models involving transport of FA from RE pathway have been partially discarded. [^{14}C]Ferulated polysaccharides were observed both in protoplasm and apoplast when wheat roots were incubated with *trans*-[U-^{14}C]cinnamic acid or L-[1-^{14}C]arabinose. The appearance of [^{14}C]ferulated polysaccharides occurred even in the presence of the lactone antibiotic brefeldin A, which is used to suppress the transport between RE and Golgi apparatus (Mastrangelo et al. 2009). The apoplastic polymerization among FA residues esterified to GAX has been related to cessation of cell wall growth and defense-related mechanisms. Ferulic acid residues polymerize after oxidation by peroxidases (Santos et al. 2008b; Oliveira et al. 2014). Oxidative coupling of polysaccharides through ester-linked feruloyl groups might occur both within the protoplast in the Golgi apparatus (dos Santos et al. 2008b; Mastrangelo et al. 2009; Umezawa 2009) and after their secretion into the apoplast. This process is catalyzed by multiple cell wall bounds and putative Golgi-resident isoperoxidases, which use hydrogen peroxide as a substrate in a mechanism similar to that of lignin polymerization (Fry 1986; Hatfield et al. 1999; Lindsay and Fry 2008).

4.6 Soluble Phenolic Compounds' Impact on Production of Cellulosic Ethanol

Unlike animals, plants lack specialized immune cells and immunological memory. However, each plant cell has developed the capability of sensing pathogens and mounting immune responses. Plant lineages have developed effective mechanisms to resist to almost all infectious agents in their environment (Jones and Dangl 2006;

Sticher et al. 1997; Thakur and Sohal 2013). The biosynthetic routes of phenolic metabolism derive from primary metabolism, and there is increasing evidence that duplications of essential genes of primary metabolism have been an important basis for gene recruitment in secondary metabolism. In the course of evolution, these duplicated genes have acquired new functions and have been optimized and diversified for their roles in new pathways (Dixon and Steele 1999). Lignin precursors themselves might exert a toxic effect on pathogens (Moura et al. 2010). In spite of the protective functions of the cinnamic acid and their hydroxylated derivatives against oxidative stress, plants themselves are not very tolerant to cinnamic acid itself. It acts as an ionophore, carrying protons from the apoplast to the protoplast through cell wall (Weng et al. 2010). In addition, cinnamic acid and hydroxycinnamates are enzyme inhibitors (Baziramakenga et al. 1995). The excess of cinnamic acid induces a strong electrolyte leakage, impairing plant nutrition and starting defensive reactions (dos Santos et al. 2008a). To top it off, cinnamic acid and derivatives show antiauxin activity that severely affects the plant growth (Baziramakenga et al. 1995; Salvador et al. 2013). Many plants exude these toxic phenols affecting the growth and development of plant competitors unable to cope with the presence of the toxic compound (Einhellig 2004; dos Santos et al. 2014).

Efficient saccharification of cell wall requires a set of enzymes such as cellulases, hemicellulases, and accessory enzymes. They act synergistically in the hydrolysis of cell wall polysaccharides. The presence of soluble phenolic compounds stored in vacuoles and apoplast and released by pretreatments impacts severely the activity of enzymes used to hydrolyze the cell wall polysaccharides (Bhat and Bhat 1997; Eriksson et al. 2002). Indeed, the presence of phenolic compounds appears to be the strongest cause of the reduction in the activity of cellulases (Ximenes et al. 2010; Vohra et al. 1980). Delignification procedures promote digestibility of cellulose by enhancing the accessibility of cellulases to cellulose and also by reducing the adsorption of enzymes to lignin (Ko et al. 2014). However, lignin degradation products can also inhibit the enzymatic hydrolysis of cellulose and reduce production of fermentable sugars.

Although understanding the inhibition of enzymatic hydrolysis by phenolic compounds is critical to the development of techniques to promote the efficiency of bioconversion of lignocellulose (Qin et al. 2016), our knowledge about the inhibition of cellulases by phenolics is still limited. Pan (2008) investigated the inhibitory effects caused by phenolic compounds and found that the presence of phenolic hydroxyl groups is critical to their inhibitory activity. He observed that phenolic hydroxyl groups could be correlated with the intensity of their inhibitory effect on cellulases. His data suggests that the interference of phenolic hydroxyl groups on the enzyme activity is more important than the steric effect and nonspecific adsorption of cellulase to lignin. Yun et al. (2014) determined the kinetics of vanillin inhibition on the cellulase, verifying that it presents competitive and reversible inhibition with the enzyme and that its aldehyde and phenolic hydroxyl groups are responsible for the inhibitory effect. More recently, Qin et al. (2016) found that different degrees of inhibition by phenolic compounds are related to the presence of different functional groups.

The group suggests that apart from the hydroxyls, the presence of methoxyl groups in the aryl groups and the level of oxidation of the carbonyl groups increase the degree of inhibition of phenolic compounds on cellulases.

4.7 Futures Perspectives

Phenolic compounds are important components in plant ecological interactions. They protect the plant from biologic, chemical, and physical attacks limiting accessibility of pathogens, herbivores, and industries to their valuable biomass. The main strategies to circumvent the negative impact of phenolic compounds on saccharification are the genetic control of the production of phenylpropanoid (Vanholme et al. 2010). Chen and Dixon (2007) downregulated the genes for C4H, HCT, C3′H, F5H, CCoAOMT, or COMT in alfalfa revealing that lines suppressed in C4H, HCT, and C3′H present the lowest lignin level (<50%) and almost twice the enzymatic saccharification efficiencies. In turn, alfalfa lines suppressed in COMT, CCoAOMT, and F5H produced little effect on lignin content and digestibility efficiencies. Li et al. (2008) reported that the impact of downregulation of genes codifying for the first enzymes of the phenylpropanoid pathway (PAL, C4H, HCT, and C3′H, Fig. 4.2) strongly reduced lignin content and biomass (Chen and Dixon 2007; Li et al. 2008; Poovaiah et al. 2014), while downregulation of F5H or COMT reduced the lignin S:G ratio, but has a smaller effect on lignin content (Li et al. 2008) or digestibility. While strong reductions in lignin content lead to an unavoidable reduction in biomass production, Jung and Phillips (2010) demonstrated a putative mutation in maize seedling that reduces the content of ferulate esters and ether cross-linking in the cell wall and increases the biomass digestibility without affecting plant growth and yield.

In addition to the genetic approach, new kinds of pretreatment have been developed to get rid of phenolic compounds before or during their production enzyme digestion. Our group has been developing an in vivo pretreatment we dubbed physiological engineering in which we apply specific inhibitors of phenylpropanoid enzymes in order to control the production of specific phenylpropanoids as ferulic and *p*-coumaric acids without reducing other relevance for production of biomass. The preliminary results suggest that small reduction in hydroxycinnamate content can strongly enhance the digestibility of grasses as signal grass, maize, and sugarcane (dos Santos et al. 2008a, b; Ita 2012; Salvador et al. 2013; Parizotto 2012; Ferro 2014). Another related approach under development consists in inducing plants to produce chemically modified lignin. At least five types of these designer lignins have been proposed: (1) lignins with a lower degree of polymerization, (2) lignins that are less hydrophobic, (3) lignins with fewer bonds to structural carbohydrates, (4) lignins containing chemically labile bonds, and (5) lignins designed to harbor value-added chemical moieties (Mottiar et al. 2016). Although much of the concrete results of this approach were obtained in in vitro assays (Tobimatsu et al. 2008, 2012; Lan et al. 2015) there are evidences that plants transformed to superexpress the production of a specific phenolic compound (Stewart et al. 2009; Wilkerson et al. 2014) or feed with phenolic compounds (dos Santos et al. 2008a, b; Salvador et al. 2013; Lima et al. 2013) incorporate them in the lignin structure.

Designer lignin has the potential to revolutionize the field both by providing plants and/or agronomical treatments to produce industrially friendly lignocellulose and by using plant metabolism as biofactories to produce value-added phenolic crude matter.

References

Anders N, Wilkinson MD, Lovegrove A, Freeman J, Tryfona T, Pellny TK, Weimar T, Mortimer JC, Stott K, Baker JM, Defoin-Platel M, Shewry PR, Dupree P, Mitchell RAC (2012) Glycosyl transferase in family 61 mediate arabinofuranosyl transfer onto xylan in grasses. PNAS 109(3):989–993

Bartley LE, Peck ML, Kim SR, Ebert B, Manisseri C, Chiniquy DM, Sykes R, Gao L, Rautengarten C, Vega-Sanchez ME, Benke PI, Canlas PE, Cao P, Brewer S, Lin F, Smith WL, Zhang X, Keasling JD, Jentoff RE, Foster SB, Zhou J, Ziebell A, An G, Scheller HV, Ronald PC (2013) Overexpression of a BAHD acyltransferase, OsAt10, alters rice cell wall hydroxycinnamic acid content and saccharification. Plant Physiol 161(4):1615–1633

Baziramakenga R, Leroux GD, Simard RR (1995) Effects of benzoic and cinnamic acids on membrane permeability of soybean roots. J Chem Ecol 21:1271–1285

Bhat MK, Bhat S (1997) Cellulose degrading enzymes and their potential industrial applications. Biotechnol Adv 15(3–4):583–620

Buanafina M (2009) Feruloylation in grasses: current and future perspectives. Mol Plant 2:861–872

Buckeridge MS, De Souza AP (2014) Breaking the "glycomic code" of cell wall polysaccharides may improve second-generation bioenergy production from biomass. Bioenergy Res 7:1065–1073

Buckeridge M, dos Santos WD, Souza A (2010) Routes for cellulosic ethanol. In: Sugarcane bioethanol, R&D for productivity and sustainability. C. LAB, Blucher

Burton RA, Fincher GB (2012) Current challenges in cell wall biology in the cereals and grasses. Front Plant Sci 3:130

Carpita N (1996) Structure and biogenesis of the cell walls of grasses. Annu Rev Plant Physiol Plant Mol Biol 47:445–476

Carpita N, Gibeaut D (1993) Structural models of primary cell walls in flowering plants: consistency of molecular structure with the physical properties of the walls during growth. Plant J 3:1–30

Carpita N, McCann M (2000) The cell wall. In: Buchanan B, Jones R (eds) Biochemistry and molecular biology of plants. American Society of Plant Physiologists, Rockville

Carpita NC, Defernez M, Findlay K, Wells B, Shoue DA, Catchpole G, Wilson RH, McCann MC (2001) Cell wall architecture of the elongating maize coleoptile. Plant Physiol 127(2):551–565

Chen F, Dixon R (2007) Lignin modification improves fermentable sugar yields for biofuel production. Nat Biotechnol 25:759–761

D'Auria JC (2006) Acyltransferase in plants: a good time to be BAHD. Curr Opin Plant Biol 9:331–340

Dixon RA, Steele CL (1999) Flavonoids and isoflavonoids: a gold mine for metabolic engineering. Trends Plant Sci 4(10):394–400 dos Santos WD, Ferrarese MLL, Nakamura CV, Mourão KSM, Mangolin CA, Ferrarese-Filho O (2008a) Soybean (Glycine max) root lignification induced by ferulic acid. The possible mode of action. J Chem Ecol 34:120–1241

dos Santos WD, Ferrarese MLL, Ferrarese-Filho O (2008b) Ferulic acid: an allelochemical troublemaker. Funct Plant Sci Biotechnol 2:47–55

dos Santos WD, Marchiosi R, Vilar FCM, Lima RB, Soares AR, Parizotto AV, Oliveira DM, Ferrarese-Filho O (2014) Polyvalent lignin: recent approaches in determination and applications. In: Lu F (ed) Structural analysis, applications in biomaterials and ecological significance, Biochemistry research trends. Nova, New York, 419 p

Einhellig FA (2004) Mode of allelochemical action of phenolic. In: Macias FA et al (eds) Allelopathy: chemistry and mode of action of allelochemicals. CRC Press, Boca Raton, pp 217–238

Eriksson T, Karlsson J, Tjerneld F (2002) A model explaining declining rate in hydrolysis of lignocellulose substrates with cellobiohydrolase I (cel7A) and endoglucanase I (cel7B) of *Trichoderma reesei*. Appl Biochem Biotechnol 101(1):41–60

Ferro AP (2014) Parede Celular de cana-de-açúcar: enzimas da via de fenilpropenoides e a biossíntese de lignina. Tese de Doutorado em Ciências Biológicas (Biologia Celular)—Universidade Estadual de Maringá

Fry S (1986) Cross-Linking of matrix polymers in the growing cell walls of angiosperms. Annu Rev Plant Physiol 37:165–186

Harris P, Trethewey J (2010) The distribution of ester-linked ferulic acid in the cell walls of angiosperms. Phytochem Rev 9:19–33

Hatfield R, Ralph J, Grabber J (1999) Cell wall cross-linking by ferulates and diferulates in grasses. J Sci Food Agric 79:403–407

Huang R, Su R, Qi W, He Z (2011) Bioconversion of lignocellulose into bioethanol process intensification and mechanism research. Bioenergy Res 4:225–245

Ita AG (2012) Efeito do ácido metilenodioxicinâmico sobre a recalcitrância do bagaço à digestão enzimática de diferentes cultivares de cana-de-açúcar. Tese de Mestrado em Agronomia. Universidade Estadual de Maringá

Jones JD, Dangl JL (2006) The plant immune system. Nature 444:323–329

Jung H, Phillips R (2010) Putative seedling ferulate ester (sfe) maize mutant: morphology, biomass yield, and stover cell wall composition and rumen degradability. Crop Sci 50:403–418

Ko JK, Ximenes E, Kim Y, Ladisch MR (2014) Adsorption of enzyme onto lignins of liquid hot water pretreated hardwoods. Biotechnol Bioeng 112(3):447–456

Lan W, Lu F, Regner M, Zhu Y, Rencoret J, Ralph SA, Zakai UI, Morreel K, Boerjan W, Ralph J (2015) Tricin, a flavonoid monomer in monocot lignification. Plant Physiol 167:1284–1295

Li X, Weng J-K, Chapple C (2008) Improvement of biomass through lignin modification. Plant J 54:569–581

Lima RB, Salvador VH, dos Santos WD, Bubna GA, Finger-Teixeira A, Marchiosi R, Ferrarese MLL, Ferrarese-Filho O (2013) Enhanced lignin monomer production caused by cinnamic acid and its hydroxylated derivatives inhibits soybean root growth. PLoS One 8:e80542

Lindsay S, Fry S (2008) Control of diferulate formation in dicotyledonous and gramineous cell-suspension cultures. Planta 227:439–452

Mastrangelo LI, Lenucci MS, Piro G, Dalessandro G (2009) Evidence for intra and extra-protoplasmic feruloylation and cross-linking in wheat seedling roots. Planta 229:343–355

Mitchell R, Dupree P, Shewry P (2007) Novel bioinformatics approach identifies candidate genes for the synthesis and feruloylation of arabinoxylan. Plant Physiol 144:43–53

Molinari HBC, Pellny TK, Freeman J, Shewry PR, Mitchell RAC (2013) Grass cell wall feruloylation: distribution of bound ferulate and candidate gene expression in Brachypodium distachyon. Front Plant Sci 4(50):1–10

Moreira-Vilar FC, Siqueira-Soares RC, Finger-Teixeira A, Oliveira DM, Ferro AP, Rocha GJ, Ferrarese MLL, dos Santos WD, Ferrarese-Filho O (2014) The acetyl bromide method is faster, simpler and presents best recovery of lignin in different herbaceous tissues than Klason Thioglycolic acid methods. PLoS One 9(10):1–7

Mottiar Y, Vanholme R, Boerjan W, Ralph J, Mansfield SD (2016) Designer lignins: harnessing the plasticity of lignification. Curr Opin Biotechnol 37:190–200

Moura JCMS, Bonine CAV, De Oliveira FVJ, Dornelas MC, Mazzafera P (2010) Abiotic and biotic stresses and changes in the lignin content and composition in plants. J Integr Plant Biol 52:360–376

Oliveira DM, Finger-Teixeira A, Mota TR, Salvador VH, Moreira-Vilar FC, Molinari HBC, Mitchell RAC, Marchiosi R, Ferrarese-Filho O, dos Santos WD (2014) Ferulic acid: a key component in grass lignocellulose recalcitrance to hydrolysis. Plant Biotechnol J 13:1224–1232

Pan X (2008) Role of functional groups in lignin inhibition of enzymatic hydrolysis of cellulose to glucose. J Biobaased Mater Bioenergy 2:25–32

Parizotto AV (2012) Cafeato o-metiltransferase de milho (Zea mays L.): Estrutura da enzima, prospecção de inibidores e recalcitrância da parede celular. Tese de Doutorado em Ciências Biológicas (Biologia Celular). Universidade Estadual de Maringá

Pedersen J, Vogel K, Funnell D (2005) Impact of Reduced Lignin on Plant Fitness. Crop Sci 45:812

Piston F, Uauy C, Fu L, Langston J, Labavitch J, Dubcovsky J (2010) Down-regulation of four putative arabinoxylan feruloyl transferase genes from family PF02458 reduces ester-linked ferulate content in rice cell walls. Planta 231:677–691

Poovaiah CR, Nageswara-Rao M, Soneji JR, Baxter HL, Stewart CNJR (2014) Altered lignin biosynthesis using biotechnology to improve lignocellulosic biofuel feedstocks. Plant Biotechnol J 12(9):1163–1173

Qin L, Li WC, Liu L, Zhu JQ, Li X, Li BZ, Yuan YJ (2016) Inhibition of lignin-derivated phenolic compounds to cellulose. Biotechnol Biofuels 9:70

Renger A, Steinhart H (2000) Ferulic acid dehydrodimers as structural elements in cereal dietary fiber. Eur Food Res Technol 211:422–428

Salvador VH, Lima RB, dos Santos WD, Soares AR, Böhm PAF, Marchiosi R, Ferrase MLL, Ferrarese-Filho O (2013) Cinnamic Acid Increases Lignin Production and Inhibits Soybean Root Growth. PLoS One 8:e69105

Stewart JJ, Akiyama T, Chapple CC, Ralph J, Mansfield SD (2009) The effects on lignin structure of overexpression of ferulate 5-hydroxylase in hybrid poplar. Plant Physiol 150:621–635

Sticher L, Mauch-Mani B, Métraux JP (1997) Systemic acquired resistance. Annu Rev Phytopathol 35:235–270

Thakur M, Sohal BS (2013) Role of elicitors in inducing resistance in plants against pathogen infection: a review. ISRN Biochem Article ID 762412, 10 p

Tobimatsu Y, Takano T, Kamitakahara H, Nakatsubo F (2008) Studies on the dehydrogenative polymerizations (DHPs) of monolignol β-glycosides: Part 4. Horseradish peroxidase-catalyzed copolymerization of isoconiferin and isosyringin. Holzforschung 62:495–500

Tobimatsu Y, Elumalai S, Grabber JH, Davidson CL, Pan X, Ralph J (2012) Hydroxycinnamate conjugates as potential monolignol replacements: *in vitro* lignification and cell wall studies with rosmarinic acid. ChemSusChem 5:676–686

Umezawa T (2009) The cinnamate/monolignol pathway. Phytochem Rev 9:1–17

Vanholme R, Ralph J, Akiyama T, Lu F, Pazo JR, Kim H, Christensen JH, Van Reusel B, Storme V, De Rycke R et al (2010) Engineering traditional monolignols out of lignin by concomitant up-regulation of F5H1 and down-regulation of COMT in Arabidopsis. Plant J 64:885–897

Vogel J (2008) Unique aspects of the grass cell wall. Curr Opin Plant Biol 11:301–307

Vohra RM, Shirsot CK, Dhawan S, Gupta KG (1980) Effect of lignin and some of its components on the production and activity of cellulase(s) by *Trichoderma reesei*. Biotechnol Bioeng 22:1497–1500

Weng J-K, Chapple C (2010) The origin and evolution of lignin biosynthesis. New Phytol 187:273–285

Weng J-K, Akiyama T, Bonawitz ND, Li X, Ralph J et al (2010) Convergent evolution of syringyl lignin biosynthesis via distinct pathways in the lycophyte Selaginella and flowering plants. Plant Cell 22:1033–1045

Wilkerson CG, Mansfield SD, Lu F, Withers S, Park J-Y, Karlen SD, Gonzales-Vigil E, Padmakshan D, Unda F, Rencoret J, Ralph J (2014) Monolignol ferulate transferase introduces chemically labile linkages into the lignin backbone. Science 344:90–93

Ximenes E, Kim Y, Mosier N, Dien B, Ladisch M (2010) Inhibition of cellulases by phenols. Enzyme Microb Technol 46:170–176

Yun L, Benkun Q, Yinhua W (2014) Inhibitory effect of vanillin on cellulose activity in hydrolysis of cellulosic biomass. Bioresour Technol 167:324–330

Zobiole L, dos Santos WD, Bonini E, Ferrarese-Filho O, Kremer R, Oliveira-Júnior R (2010) Lignin: from nature to industry. In: Paterson R (ed) Lignin: properties and applications in biotechnology and bioenergy. Hauppauge, New York, pp 419–436

Part II
Microbial Enzymes

Chapter 5
Enzymes Involved in the Biodegradation of Sugarcane Biomass: Challenges and Perspectives

Maria de Lourdes T.M. Polizeli, Alexandre Favarin Somera, Rosymar Coutinho de Lucas, Monica Stropa Ferreira Nozawa, and Michele Michelin

Abstract This chapter introduces the role of enzymes in the biomass degradation, namely sugarcane bagasse and straw, leading to the formation of fermentable sugars and second-generation ethanol. The chapter begins with a retrospective of the structuring of the ethanol production chain and then presents current aspects where the deficit of production and its consequences in business can be seen. Subsequently, we list the role of enzymes involved in the biomass hydrolysis, the commercial cocktails, and the proposal of our laboratory in this context. On the other hand, the efficiency of enzymes on the biomass is increased when the bagasse and straw are pretreated. Thus, some technologies that may facilitate the enzymatic hydrolysis and the formation of fermentable sugars are described. Lastly, recent analytical methods that enable a better analysis of the composition and viewing of fiber in the sugarcane biomass are presented.

Keywords Sugarcane bagasse • Bioethanol • Cellulose • Hemicellulases • Lignin • Pretreatment • Analytical methods

---

M. de Lourdes T.M. Polizeli (✉) • A.F. Somera • M.S.F. Nozawa
Faculdade de Filosofia, Departamento de Biologia, Ciências e Letras de Ribeirão Preto, Universidade de São Paulo, São Paulo, Brazil
e-mail: polizeli@ffclrp.usp.br

R.C. de Lucas
Faculdade de Filosofia, Departamento de Biologia, Ciências e Letras de Ribeirão Preto, Universidade de São Paulo, São Paulo, Brazil

Faculdade de Medicina de Ribeirão Preto, Departamento de Bioquímica e Imunologia, Universidade de São Paulo, São Paulo, Brazil

M. Michelin
Faculdade de Filosofia, Departamento de Biologia, Ciências e Letras de Ribeirão Preto, Universidade de São Paulo, São Paulo, Brazil

Centro de Engenharia Biológica, Universidade do Minho, Braga, Portugal

M.S. Buckeridge, A.P. De Souza (eds.), *Advances of Basic Science for Second Generation Bioethanol from Sugarcane*, DOI 10.1007/978-3-319-49826-3_5

5.1 Introduction

This chapter provides an overview of the production of second-generation ethanol regarding the enzymes that degrade the cell walls of sugarcane. Also, there is an overview of the pretreatments that can be carried out on the straw and bagasse of sugarcane, and the current techniques that could further elucidate details of the structure of the lignocellulolytic biomass. In this context, initially, a retrospective and the actual scenario of production of the sugarcane in Brazil is described, since it is one of the main plant species cultivated in this country, and the Southeastern and Midwestern regions are the most important areas for it. After that, this chapter describes how sugarcane by-products such as straw and bagasse could be used to enhance the production of bioethanol.

Despite many advances in the area of enzymology related to bioenergy, it should be remembered that in order to maintain the efficiency of converting cellulose and hemicellulose into glucose and xylose, it is necessary to (a) reduce the cost of the enzyme production, (b) use nonpathogenic microorganisms, (c) pretreat biomass, (d) avoid the inhibition of the enzymes by their own products of the assay, and (e) not allow interferences of the lignin degradation products in the microorganism growth. Additionally, a detailed description of the action of each enzyme involved in the hydrolysis of the sugarcane cell wall, the efficiency of commercial enzymatic cocktails and the goals of our laboratory are presented.

Furthermore, a section with a brief explanation of some of the pretreatment technologies used with sugarcane bagasse, in order to improve the access of hydrolytic enzymes in the macromolecules of the cell wall, is presented. It includes some physical, physicochemical, chemical, and biological pretreatments. Finally, in the last section, some methods for characterization of sugarcane bagasse, depolymerase activity, hydrolysis products, and inhibitors are described. Alternative analyses of cellulose and hemicellulose hydrolysis products are addressed and a synergistic view of the hydrolytic enzymatic capacities on sugarcane cell wall and the methodologies of analysis of biomass are discussed.

5.2 Retrospective and Current Deficit of Ethanol Production

The sugarcane crops were brought to Brazil in 1532 and the cultivation of this crop was the economic basis for the increase in population, mainly in the State of São Paulo and in Northeastern Brazil by Portuguese and the Dutch immigrants, respectively. According to Miranda (2010), scientific and detailed studies indicate that during the period of the Portuguese royalty, the average of annually cultivated area of sugarcane was 9000 ha, reaching as much as 16,000 ha at the beginning of the Empire. The total deforestation for the production of sugar for more than 220 years, during the years of struggle for the independence from Portugal, was around

140,000 ha. Currently, in just 1 year, Brazil produces more sugarcane than it did during the whole Imperial Age.

The advent in the 1970s of the Federal Program entitled "Pro-Alcohol" placed Brazil in the vanguard of the world scenario in the production of renewable fuels. This program led to the establishment of a large number of industrial bioethanol plants based on the sugarcane industry already established with existing sugarcane-processing mills (De Souza et al. 2014), that was a time when an energy crisis was lived, which was intensified in the conferences on the environment (Soccol et al. 2010). Although Brazil has invested in the bioethanol production from sugarcane, it is still based exclusively on the "first-generation technologies," where their main product, e.g., sucrose, is subjected to the fermentation process. However, this source of energy is only 1/3 of the energy content of sugarcane, and much more bioethanol can be produced if all energy from biomass is utilized.

Although with great potential, the sugar and alcohol sector suffered a strong decline over the recent years, which is causing a significant deficit of bioethanol in the fuel market. Since 2008, the Brazilian ethanol market is showing a growing disconnection between the actual supply and the potential demand for this product. On the demand side, biofuel car sales, according to ANFAVEA (n.d.) (National Association of Automobile Manufacturers), increased by about 11% per year between 2008 and 2010 (BNDES 2015). As for the ethanol, the supply was almost stagnant during the same period. While in 2008 27.1 billion liters of the fuel were produced, and in 2010 27.9 billion liters were produced, which means an increase of only 1.5% per year (BNDES 2015). Figure 5.1 illustrates the timeline of ethanol production since 1970 until 2012.

Currently, data available from Brazilian Sugarcane Industry Association (Unica 2015) shows that the quantity of sugarcane processed in 2015/2016 referents at harvest in the South-Central region of Brazil reached 620,830 thousand tons and the ethanol production totaled 28,363 thousand m^3—with 17,692 thousand m^3 of anhydrous ethanol and 10,671 thousand m^3 of hydrous ethanol.

Bearing in mind the predicted rising in bioethanol demand due to the increase of flexible-fuel vehicles, and the fact that it is necessary to avoid the competition for land with food crops, i.e., rice, beans, corn, soybeans, coffee, or pastures, it is plausible to state that rational alternatives, incentives, and accurate public policies are necessary to increase the bioethanol production from sugarcane in Brazil. The current global trend is to use bagasse from sugarcane not only for the production of heat and the generation of the power necessary for the production process of bioethanol from sugarcane, the factor that makes a positive energy balance in the ethanol production, but also as an energy source for the second-generation bioethanol production. In Brazil, this strategy would help to supply, or at least to mitigate the bioethanol production deficit trends that have been observed in the last 8 years.

Brazil is a country with a great potential for the generation of renewable resources, which places it in a privileged position to develop technologies related to the use of biomass. The wide area with fertile soils, the intense solar radiation, the abundant water supply, and the diversity of weather and soil (Bon et al. 2008) are factors that favor Brazil to produce biomass. Thus, not only sugarcane bagasse and

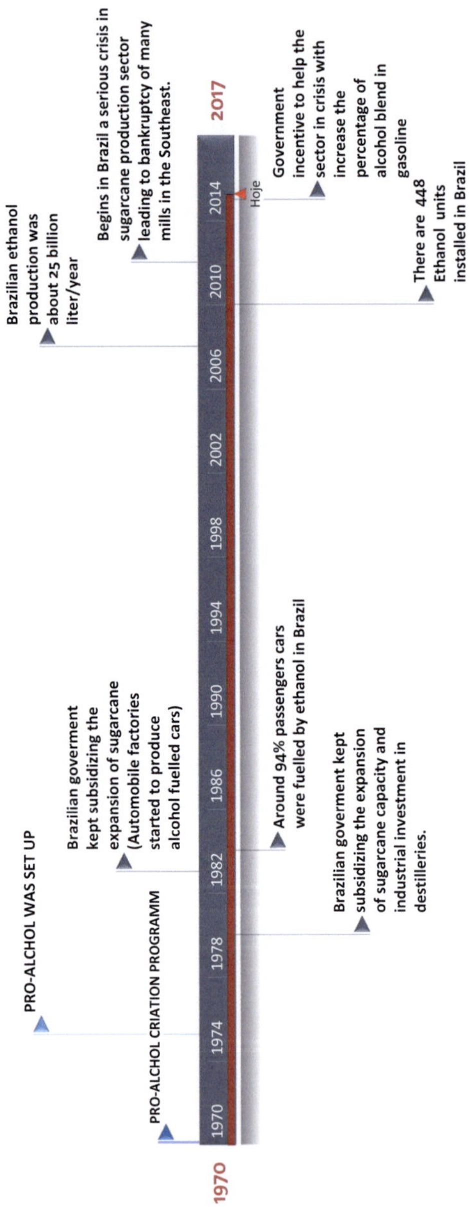

Fig. 5.1 Timeline showing the principal landmarks in the history of ethanol production in Brazil from 1970 to 2012

straw, but also other biomasses such as agro-industrial and forest residues might be used for "second-generation" biofuel production in this country.

Regarding the generation of biomass as sugarcane bagasse, it is important to consider that between 2010 and 2011 Brazil harvested about 200 million tons of bagasse as residue (CONAB 2011, cited in Rocha 2011). This material has been used mainly in burning processes for power generation within the sugar and ethanol industry. However, this product has been pursued to increase the production of alcohol fuel through the degradation of its cell wall, which is rich in complex polysaccharides that can be hydrolyzed and fermented into ethanol. However, for this process to be effective, accurate studies of its cell wall structure/organization are necessary, as well as studies to understand the main cell wall deconstruction mechanisms involved in the bioethanol production (Santos et al. 2012; De Souza et al. 2013).

The advancement and the success of second-generation bioethanol industries are highly dependent on the enzyme production. Although there is plenty of research involved in the production of enzymes from microorganisms that have potential application in the production of bioethanol, the enzyme production remains as one of the major bottlenecks to overcome cell wall recalcitrance.

5.3 Enzymatic Hydrolysis of Sugarcane Cell Wall: Cellulases and Hemicellulases—Concepts and Updates

Biomass is defined as the renewable resource of all organic material (plant or animal) that can be used for energy production. The most abundant biomass in nature is lignocellulosic materials, which are composed of agro-industrial materials, municipal waste, and the wood of angiosperms and gymnosperms (Segato et al. 2014). The main sources of lignocellulosic residues produced by the global agribusiness are sugarcane, corn, rice, and wheat (De Souza et al. 2013).

Cell walls of straw and bagasse of sugarcane are composed of a recalcitrant polymeric structure with 31–48.6% cellulose, 25–32% hemicellulose, 17–24% lignin, and 1.2–7% ashes/extractives, as the main components (De Souza et al. 2013). Cellulose and hemicellulose (represented mainly by xylan) must be degraded into glucose and xylose, respectively, using acid treatment or enzymatic hydrolysis. After that, these monomeric sugars are fermented by yeast to ethanol. Acid treatment results in several problems such as the corrosion of materials used in the process and the presence of acids in the effluents. The enzymatic hydrolysis of lignocellulosic biomass aiming at the production of fermentable sugars is the best alternative, but it is also considered as a bottleneck during the production of bioethanol.

Based on the evolutionary expression of cell wall formation, Buckeridge and De Souza (2014) suggested the inclusion of cell wall polysaccharide as part of sugar code, named for them by Glycomic Code. This code is the assumption that the composition and function of cell walls in plant tissues and organs are derived from the communication mechanism between biosynthesis and degradation of polysaccharides.

In addition, they suggest that different combinations of polymers with unique fine structures arrange themselves within the wall creating a code. In this way, comprehending the Glycomic Code will enable the emergence of new hydrolysis technologies for the plant cell wall polysaccharide degradation (Buckeridge and De Souza 2014; Tavares and Buckeridge 2015; Buckeridge et al. 2015).

Because of the complexity and recalcitrance of cell walls, the major difficulty in the saccharification process is related to hydrolysis. To be efficient, the use of lignocellulosic material to cellulosic ethanol production requires a controlled hydrolysis of all insoluble polysaccharides present in this biomass. In addition, to enhance the biomass saccharification, the cellulosic ethanol industry has been using a variety of strategies, such as physical and chemical pretreatments (Soccol et al. 2010). Although these processes promote the lignin extraction, they might increase the cellulose crystallinity, hindering the saccharification process (Driemeier et al. 2011). Only after the lignin removal and elimination of the enzymatic inhibitors, such as furfural and hydroxymethylfurfural, biomass is ready for the enzymatic saccharification.

Some of the most important enzymes applied in this process are the cellulases and hemicellulases, i.e., hydrolases working synergistically to release sugars. They attack the remaining polysaccharides producing sugars (C5 or C6) that can be directly converted into ethanol by yeasts in the fermentation process (Castro and Pereira 2010; Hu et al. 2013).

5.3.1 Cellulases

Cellulose is the major structural constituent of plant cell wall and it is composed by the linear homopolysaccharide formed by D-glucose units joined by glycosidic bonds of the β-1,4 type. Its role is structural, providing mechanical protection to the cell (Segato et al. 2014).

The enzymes responsible for the cellulose breakdown are cellulases, a group of hydrolases that cleave β-1,4 O-glycosidic bonds. They are classified into three groups, according to the site where they act on their substrate molecules: (a) endoglucanases (EnG) or endo-β-1,4-glucanases (EC 3.2.1.4), which cleave the internal bonds of cellulose fibers present in the amorphous part of cellulose, releasing cello-oligosaccharides; (b) exoglucanases (ExG), or cellobiohydrolases (CBH) (EC 3.2.1.91, EC 3.2.1.176), which act on the external region of cellulose, releasing cellobiose from the reducing- (CBH I) or nonreducing ends (CBHII); and (c) β-glucosidases or cellobiases (EC 3.2.1.21) hydrolyze soluble oligosaccharides and cellobiose up to glucose (Zhang et al. 2006; Castro and Pereira 2010). Figure 5.2 (first part) illustrates the action mechanism of the main cellulases by showing the action of enzymes on cellulose.

Two catalytic residues are responsible for the hydrolysis: a proton donor (generally acid) and a nucleophile/base. Hydrolysis occurs depending on the residue spatial catalytic position, either by the retention or the inversion of the anomeric configuration. Cellulases retain the same configuration after a double-displacement

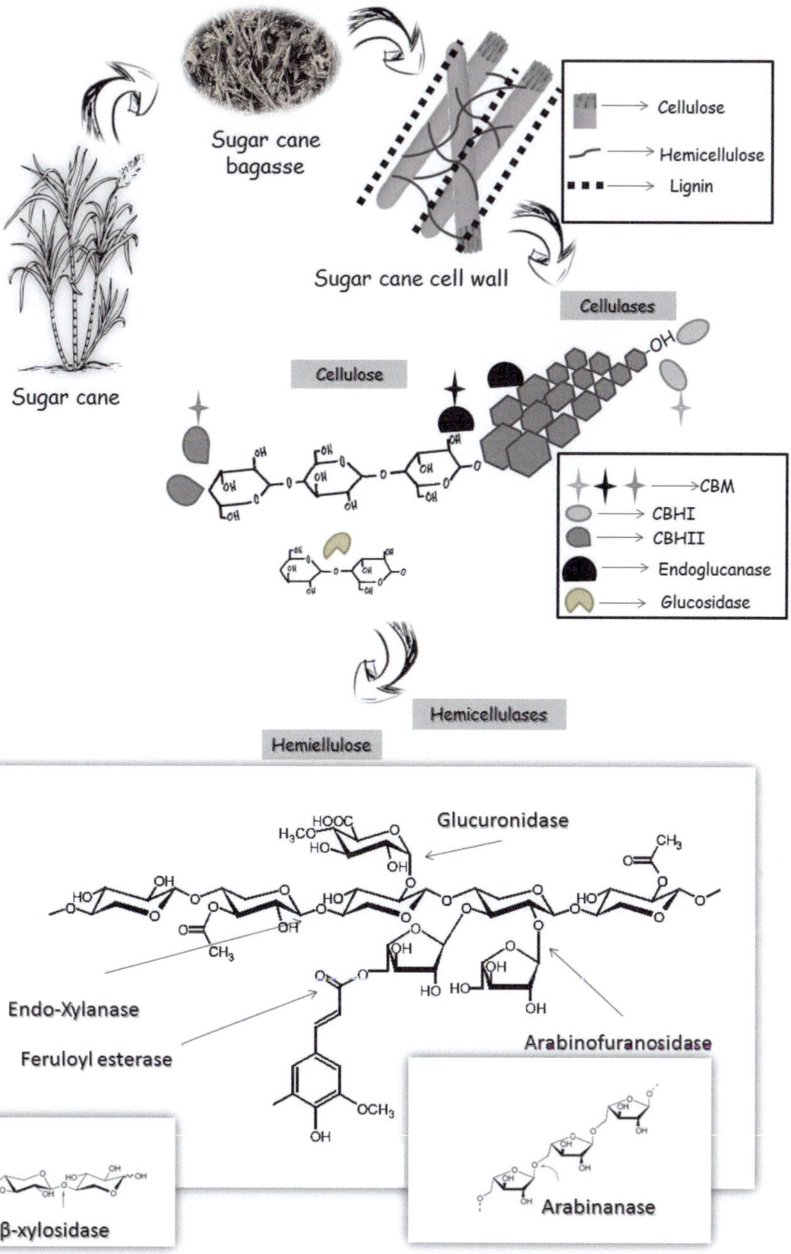

Fig. 5.2 Action mechanism of the main holocellulolytic enzymes. The second part is an adaptation of Dodd and Cann (2009)

hydrolysis with two glycosylation/deglycosylation steps of the anomeric C bond. On the other hand, the "inverting" cellulases invert their configuration after a single nucleophilic displacement hydrolysis (Zhang et al. 2006; Payne et al. 2015). Thus, different action mechanisms of this group of enzymes enhance catalytic efficiency with different substrates increasing the hydrolytic properties and thereby promoting the conversion of oligomers to be used in the fermentative process for the production of second-generation ethanol (Vuong and Wilson 2010).

In order to efficiently deconstruct cellulose polymer, cellulolytic enzymes attack the substrate in a synergistic way. This synergism results in an increase of efficiency by the action of two or more components of the system as compared to the action of each one separately.

Transmission electron microscopy, X-ray diffraction, and atomic force microscopy techniques have been used to determine the three-dimensional structure of cellobiohydrolases. The results showed that these enzymes possess two distinct domains, one catalytic domain and one carbohydrate-binding domain (CBD) linked by a flexible region rich in threonine, proline, or serine residues, called hinge (Gilkes et al. 1991). Although the CBD is not essential for the catalytic activity of the enzyme, it modulates the specific enzymatic activity in soluble and insoluble cellulolytic substrates. CBD binds on crystalline cellulose, weakening the hydrogen bonds among microfibrils, helping in their separation and favoring the enzyme progression (Segato et al. 2012; Pakarinen et al. 2014).

Another group of enzymes closely related to the cellulases is the lytic polysaccharide monooxygenases (LPMO) (GH61/CBM33), copper-dependent enzymes recently reclassified as AA9 (fungal enzymes) and AA10 (bacterial enzymes). LPMOs act on the cellulose backbone using an oxidative mechanism, improving the degradation of this recalcitrant polymer into oligosaccharides, and consequently bursting the cellulosic ethanol production. The exact mechanism of action of these enzymes is unclear, but some studies assume the interaction between the flat face of the active site and the substrate, with copper acting as the cofactor of the oxidation reaction (Hemsworth et al. 2013, 2014; Kjaergaard et al. 2014; Lo Leggio et al. 2015).

5.3.2 *Hemicellulases*

Hemicellulose is a group of a heterogeneous polymers organized into two main cores, i.e., the backbone and branches. This complex can be formed by hexoses (β-D-glucose, β-D-mannose and α-D-galactose), pentoses (β-D-xylose and α-L-arabinose), and some acid sugars (α-D-glucuronic, α-D-4-O-methyl-galacturonic, and α-D-galacturonic acid) (Gírio et al. 2010).

The variable structure and organization of hemicellulose require the action of many enzymes for its complete deconstruction such as xylanases, arabinanases, arabinofuranosidases, ferulic and coumaric acid esterases, acetyl xylan esterases, glucuronidases, mannanases, xylosidases, galactosidases, and others (De Vries and Visser 2001; Polizeli et al. 2005; Gírio et al. 2010).

Segato et al. (2014), in a recent review, suggested that hemicellulases can be divided into three groups: (1) enzymes that hydrolyze backbones; (2) enzymes that remove side chains; (3) accessory enzymes that remove substituted residues, such as acetyl groups. A detail explanation about these enzymes is described below.

1. *Enzymes that hydrolyze backbones*:

   (a) *Endo-1,4-β-xylanases (EC 3.2.1.8)*: These are responsible for cleaving internal glycosidic linkages of xylan backbone, initially forming xylooligosaccharides, or forming xylotriose, xylobiose, and xylose in further steps. Two families belong to this group of enzymes, i.e., GH10 and GH11. While the GH10 family is able to hydrolyze xylan chains closer to the branching points, the GH11 family is able to cleave this polymer in more distant regions of the side chains (Dodd and Cann 2009; Pollet et al. 2010). Thus, branching residue distribution throughout the polymer seems to interfere in the action of these endo-cleaving enzymes on the xylan backbone (Polizeli et al. 2005; Nacke et al. 2012).
   (b) *β-D-xylosidases: (EC 3.2.1.37)*: These enzymes hydrolyze small xylooligosaccharides and xylobiose from their nonreducing ends, releasing free xylose. They are usually not able to hydrolyze xylan, but can cleave artificial substrates such as *p*-nitrophenyl β-D-xylopyranoside (Kurakabe et al. 1997; Polizeli et al. 2005).
   (c) *Mannanase (EC 3.2.1.78)*: They act on the galacto (gluco) mannan skeleton, releasing predominantly mannobiose and mannotriose (Buckeridge 2010; Segato et al. 2014).
   (d) *β-Mannosidases (EC 3.2.1.25)*: These enzymes are exo-hydrolases working on nonreducing ends of mannan-oligosaccharides (MOS), releasing mannose (Buckeridge 2010; Segato et al. 2014).

2. *Enzymes that remove side chains*:

   (a) *Arabinanases (EC 3.2.1.99)*: They act on the linear arabinan. They do not exhibit activity on the synthetic substrate *p*-nitrophenyl α-L-arabinofuranoside (Segato et al. 2014).
   (b) *Arabinofuranosidases (EC 3.2.1.55)*: These act on the branched arabinans, removing residues substituted at positions 2 and 3 of xylan. They are active on the synthetic substrate *p*-nitrophenyl α-L-arabinofuranoside (Gírio et al. 2010; Segato et al. 2014, see Chap. 6).
   (c) *Xyloglucanases (EC 3.2.1.151)*: These enzymes are able to hydrolyze xyloglucan backbone (glucose β-1,4 linkages) in glucose residues (Buckeridge 2010; Segato et al. 2014).
   (d) *α-Glucuronidases (EC 3.2.1.139)*: They hydrolyze the α-1,2 linkages between glucuronic acid and xylose residues in glucuronoxylan (Tenkanen and Siika-aho 2000; Gírio et al. 2010).
   (e) *α-Galactosidase (EC 3.2.1.22) and β-galactosidase (EC 3.2.1.23)*: These are enzymes acting on D-galactose residues. β-Galactosidase releases galactose from side chains of pectin galactan, whereas the α-galactosidase removes galactose from mannose residues in the galacto (gluco) mannan skeleton (Buckeridge et al. 2000; Gírio et al. 2010).

3. *Accessory enzymes that remove substituted residues, such as acetyl groups*:
   (a) *Acetyl xylan esterases (EC 3.1.1.72)*: These remove *O*-acetyl substituents from the position C2 and/or C3 of xylose residues in acetyl xylan (Caufrier et al. 2003; Polizeli et al. 2005). They are important enzymes involved in the xylan saccharification, since they facilitate the endoxylanase action on the xylan backbone (Gírio et al. 2010; Segato et al. 2014, see Chap. 6).
   (b) *Feruloyl esterases (EC 3.1.1.73) and p-coumaric acid esterases (EC 3.1.1.)*: These are esterases able to cleave ester linkages in xylan, between the arabinose side chains and ferulic acid, and from arabinose and *p*-coumaric acid, respectively (Crepin et al. 2004; Polizeli et al. 2005; Gírio et al. 2010; Segato et al. 2014). Figure 5.2 (second part) illustrates the action mechanism of the main hemicellulases by showing the action of enzymes on arabinoxylan.

Due to the insolubility and recalcitrance of plant cell wall, many hemicellulases are modular proteins, and besides their catalytic domain, they possess other functional modules. The catalytic modules present in the hemicellulases are glycoside hydrolase (GHs). They are responsible for the hydrolysis of glycosidic linkages. Another domain is the carbohydrate esterase (ECs), which hydrolyzes ester bonds of acetate groups or ferulic acid. These modules are constructed based on the homology of their primary sequence, conservation of tertiary and secondary structure, and molecular mechanisms of catalysis (Henrissat and Davies 2000; Cantarel et al. 2009).

5.3.3 Commercial Enzymes: Current State of the Art

The most important step in producing cellulosic ethanol is the enzymatic hydrolysis of the biomass, which aggregates high cost to the saccharification process (De Souza et al. 2013; Gírio et al. 2010). Thus, several strategies to decrease the enzyme cost in this process have been adopted over the years. Scientists around the world have been developing approaches to improve enzyme performance to make the cellulosic ethanol industry worthy. In this scenario, one of the most important procedures is the prospection of enzymes capable of performing the deconstruction of plant cell walls. The GH61 or LPMO is an example of how the introduction of new enzymes in the current commercial cocktails might help to increase the efficiency of cell wall hydrolysis. Many experiments with this enzyme have shown that it can increase the cellulose conversion yield, and reduce the amounts of other enzymes necessary to saccharify biomass (Cannella et al. 2012).

The final goal for the production of cellulosic ethanol is the conversion of polymers in monomers especially by glucosidases. Nonetheless, this type of enzyme is usually inhibited by the substrate. Thus, the prospection of glucosidases resistant to the substrate inhibition, or their genetic manipulation, is a sine qua non step for the saccharification process (Rani et al. 2014).

In the past years, commercial enzyme companies have made significant progress in producing new-generation cocktails with enzymes presenting higher specific activities and lower cost using different biotechnology processes and engineering approaches, especially after signing a contract with the US Department of Energy (DOE), whose main goal was to decrease the enzymatic production cost approximately 20-fold (McMillan et al. 2011). Since then, Novozymes and Genencor have been working in launching efficient enzymatic preparations to improve the sugar conversion. Thus, the newest generations of cocktails that can be cited are the Cellic Ctec® and Htec® family from Novozymes, and Accellerase from Genencor (de Paula et al. 2016; Kallioinen 2014; DuPont 2013; Novozymes 2012).

Cellic Ctec 3®, the last family launched by Novozymes, is a cellulose and hemicellulase complex able to convert pretreated lignocellulosic materials into fermentable sugars. In addition, Htec3 from the same company is only a hemicellulase complex that can supplement other cocktails aiming at the deconstruction of biomasses rich in hemicellulose. On the other hand, Accellerase 1500®, the last complex launched by Genencor, is an enzyme complex that contains exoglucanase, endoglucanase, hemicellulase, and β-glucosidase, which can also work in a variety of pretreated biomass (de Lucas et al. 2016; de Paula et al. 2016; Li et al. 2015; DuPont 2013; Canella et al. 2012; Novozymes 2012).

The large effort of industries around the world working in decreasing the enzymatic production costs, elevating the enzymatic efficiency, generating a large amount of sugars, and consequently producing ethanol from biomass is also a big step to reduce the impact caused by residues constantly thrown in nature, aggregating value to this product (Fig. 5.3).

In this perspective, scientists around the world have been really contributing with studies on the elucidation of the enzymatic way of action involved in biomass deconstruction and the discovering of new enzymes and new potential microorganisms. In addition, with recent methodology sources such as molecular biology, they are developing great catalysts that can be used in enzymatic cocktails applied in plant cell wall hydrolysis.

In Brazil, the Laboratory of Microbiology and Cell Biology from the São Paulo University, in Ribeirao Preto city, coordinated by Dr. Polizeli, has reached significant advances in the enzymology field. This group is responsible for prospecting more than 1000 new environmental filamentous fungi that can demonstrate high enzymatic secretion potential, especially enzymes applied in industrial processes. For that, they run a rigorous screening of their samples and also perform biochemical characterization of their catalysts, such as determination of temperature and pH effects on stability and enzyme activity, effect of glycosylation, determination of kinetic constants, mass spectrometry analysis, molecular modeling, immobilization, and effect of salts, inhibitors, solvents, chelate, and reducing composts on enzyme activity.

Recently, using protein heterologous expression and enzymatic immobilization on different chemical supports, Dr. Polizeli's group is trying to improve the performance of the most promising enzymes that were already screened. Another methodology recently applied by our research group is the secretome, an important

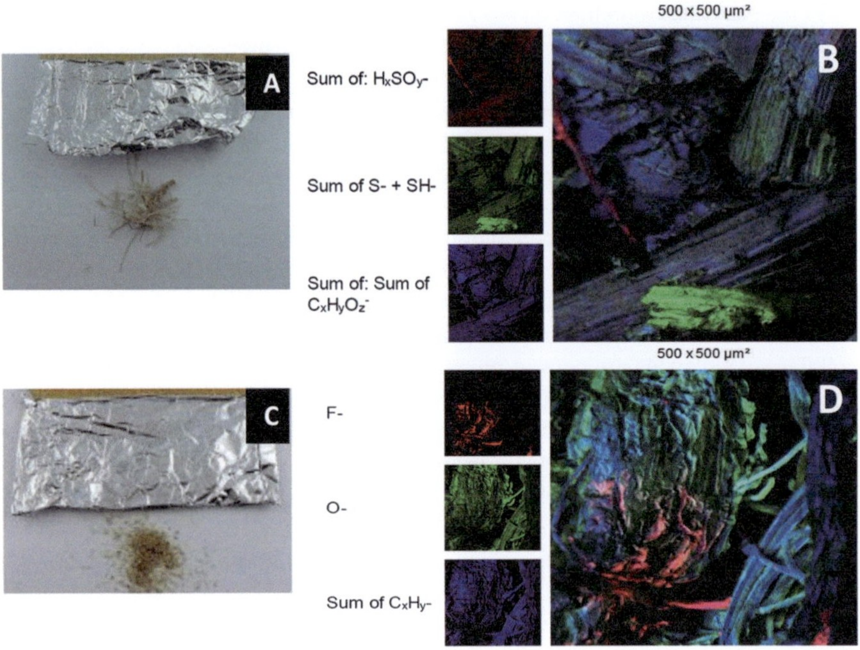

Fig. 5.3 Surface chemical imaging of sugarcane bagasse. The sample was introduced directly into the TOF-SIMS instrument without further preparation. The analysis was conducted using a 30 keV Bi3+ primary ion beam scanned over a 500 × 500 μm raster upon the surface for a measurement time of 100 s. Secondary ions emitted from the sample surface were analyzed according to mass and elemental/molecular ion images produced by synchronization with the primary ion beam position. (**a**) Sugarcane sample *in natura* with aluminum packaging. (**b**) TOF-SIMS ion image showing sulfur (*green*) distribution upon the treated sugarcane surface. (**c**) Milled sugarcane sample *in natura* with aluminum packaging. (**d**) TOF-SIMS ion image showing the presence of fluorine (*red*) distribution upon the treated sugarcane surface. Illustrations obtained from TOF.SIMS 5–100 instrument operation. The pictures were obtained from TOF.SIMS 5–100 instrument operation. We thank the efforts of Dr. Reinhard Kersting, at TASCON GmbH, Münster, Germany, and Fellipy Ferreira, at dpUNION Instrumentação Analítica e Científica Ltda, São Paulo, Brazil

source to better understand the enzymatic requirement by the fungi according to a different kind of biomass employed as its growing source. Using this methodology, it might be possible to explain, for instance, how enzymes work together and which enzymes are necessary to deconstruct the plant cell wall. Also, the information obtained from secretome can help to construct an efficient enzymatic cocktail directed to a specific biomass such as sugarcane. Borin et al. (2015) studying the secretome from *Aspergillus niger* and *Trichoderma reesei* verified that these fungi use different enzymatic sequences to degrade the same biomass, which corroborate with the hypothesis already explained here that for the construction of an efficient

Table 5.1 Recent advances from Dr. Polizeli's group on potential recombinant enzymes applied on biomass hydrolysis

Gene source	Enzymes[a]	Reference
Aspergillus niveus	Endo-β-1,4-glucanase	Furtado et al. (2015)
Malbranchea pulchella	Xylanase GH10	Ribeiro et al. (2014)
A. niveus	Endo-xylanase and α-L-arabinofuranosidase	Damasio et al. (2013)
A. niveus	Endo-β-1,4-glucanase (GH12)	Damasio et al. (2012c)
A. niveus	GH7-Cellobiohydrolases	Segato et al. (2012)
A. niveus	L-Arabinofuranosidase	Damasio et al. (2012a)
A. niveus	Endo-1,5-arabinanase	Damasio et al. (2012b)
A. niveus	Xylanase GH11	Damasio et al. (2011)
A. terreus	Xyloglucanase	Vitcosque et al. (2016)

[a]The mutant *Aspergillus nidulans* A773 was used as a host for the enzyme secretion

enzymatic cocktail it is important to verify how these fungi act when grown on different types of biomass, which kind of enzymes they produce, and, the most important, which sequence of these catalysts are necessary to deconstruct the biomass.

Table 5.1 shows some of the recent advances from Dr. Polizeli's group in the production of enzymes involved in the hydrolysis of holocellulosic biomass and their contribution to the global scientific community. With regard to the second-generation ethanol, many of these enzymes mentioned in Table 5.1 are now being tested in order to evaluate their ability in converting biomass into oligomers, and, consequently, the publications are still ongoing and the results achieved until now are promising.

5.4 Pretreatment Technologies

The recalcitrant nature of plant cell walls makes the pretreatment of lignocellulosic biomass essential for efficient and cost-effective saccharification of their macromolecules to fermentable sugars by cellulolytic and hemicellulolytic enzymes (Soccol et al. 2010).

Pretreatment technologies can be categorized as physical, physicochemical, chemical, or biological and more information about these treatments is described in Table 5.2. The goals of all pretreatment technologies are to improve the access of hydrolytic enzymes to cell wall polysaccharides while minimizing degradation of sugars and formation of fermentation inhibitors. However, lignocellulose pretreatments have a detrimental effect of releasing a wide range of compounds such as phenolics, weak acids, and furan derivatives, which are inhibitory to fermenting microorganisms and/or cellulolytic enzymes.

Table 5.2 Summary of the advantages and disadvantages of several pretreatment technologies of lignocellulosic biomass

Pretreatment	Advantages	Disadvantages
Mechanical comminution	Reduces cellulose crystallinity	High power and energy requirement
Steam explosion	Causes hemicellulose solubilization and lignin transformation; cost effective	Destruction of a portion of the xylan; incomplete disruption of the lignin-carbohydrate matrix; generation of compounds inhibitory
Liquid hot water (LHW)	Causes hemicellulose solubilization and increases the cellulose digestibility by enzymes; low or no inhibitor formation; no need for catalyst; cheap reactor systems	High energy/water input
Ammonia fiber explosion (AFEX)	Increases accessible surface area, removes lignin and hemicellulose; low formation of inhibitors	Not efficient for biomass with high lignin content; high cost of ammonia; requires recycling of the ammonia
CO_2 explosion	Increases accessible surface area; cost effective; no inhibitor formation	Does not modify lignin; very high pressure requirements
Ozonolysis	Reduces lignin content; does not produce toxic residues; operation at ambient conditions	High cost of large amount of ozone required
Acid hydrolysis	High glucose yield; hemicellulose and lignin solubilization; makes the cellulose more accessible for the enzymes	High cost of acid and need for recovery; washing and/or detoxification step requirement before the fermentation; equipment corrosion problems and requirement of more expensive reactors; inhibitor formation
Alkaline hydrolysis	Causes lignin solubilization; increases accessible surface area	High cost of alkaline catalyst; long residence time requirement; irrecoverable salts formed and incorporated into biomass; need of neutralization of the pH and/or removal of the lignin and inhibitors
Organosolv	Hydrolyzes lignin and hemicelluloses	Need of solvents to be drained from the reactor, evaporated, condensed, and recycled; high cost

Green solvents	Hydrolyzes lignin and hemicellulose; dissolves high loadings of different biomass types; mild processing conditions (low temperatures)	High solvent costs; need of solvent recovery and recycle
Pulsed electrical field	Ambient conditions; disrupts plant cells; simple equipment	Need for more research
Wet oxidation	Efficient removal of lignin, dissolves hemicellulose and causes cellulose decrystallization; low inhibitor formation; minimizes the energy demand (exothermic)	High cost of oxygen and alkaline catalyst
Ammonia recycle percolation	High efficiency for delignification; high cellulose content after pretreatment	High energy costs and liquid loading
Supercritical fluid	Increases cellulose accessible area; low degradation of sugars; cost effective	High pressure requirements; lignin and hemicelluloses unaffected
Biological	Degrades lignin and hemicelluloses; low capital cost; low energy requirements; no chemical requirement; mild environmental conditions	Low hydrolysis rate and long pretreatment times

Adapted from Kumar et al. (2009), Alvira et al. (2010), Brodeur et al. (2011), Menon and Rao (2012), Michelin et al. (2014)

5.5 Inhibition and Deactivation of Cellulases: Challenges and Perspectives

Enzymatic hydrolysis of lignocellulosic biomass for the production of fermentable sugars has been extensively investigated, being reported as a critical step during the production of biofuels and other bio-based products (Zhang and Lynd 2004; Lynd et al. 2008). As mentioned before this step represents the major contributor to the total cost of producing ethanol from biomass. Therefore, it is critical to minimize the use of enzymes while maintaining the efficiency of converting cellulose to glucose through enzymatic hydrolysis.

The cost of enzymes coupled with the large amounts required to obtain commercially viable yields is by far the main economic barrier posed to the large-scale production of biofuels. A significant contributor to the high-dose demands is that hydrolysis rates decrease when reaction proceeds in a much faster way than can be explained by substrate consumption alone for a typical enzymatic hydrolysis (Yang et al. 2006). Several factors, including enzyme inhibition and deactivation, a drop in substrate reactivity, or nonproductive binding of enzyme to lignin, could be responsible for the loss of enzyme effectiveness with time (Eriksson et al. 2002; Qing et al. 2010).

Lignin-derived inhibitors, frequently released during the pretreatment steps, are a major obstacle because they prevent the enzymatic hydrolysis of cell wall polysaccharides, especially with softwood lignocellulosic materials. However, the exact mechanisms of lignin-derived inhibition are not fully understood. It is known that lignin physically hinders the access of enzymes to cellulose (Donohoe et al. 2008; Ishizawa et al. 2009; Rollin et al. 2011; Rahikainen et al. 2011) due to adsorption of cellulase to lignin (Berlin et al. 2006; Nakagame et al. 2010), and also inhibits or deactivates cellulases due to the presence of phenols produced from lignin degradation (Jing et al. 2009; Ximenes et al. 2010, 2011).

Several works have studied the effect of these inhibitors on the enzymatic activity. For example, Ximenes et al. (2010, 2011, see Chap. 6), and Kim et al. (2011, 2013), identified phenolic compounds from lignin as the major enzyme inhibitors and/or deactivators. The amount and type of phenolic compounds depend on the biomass source, and methods and conditions of pretreatment used. The strength of the inhibition or deactivation effect, however, depends mainly on three factors: (1) the type of enzyme; (2) the microorganism from which the enzyme was derived, and (3) the type of phenolic compounds.

The inhibitory effect of several phenolic compounds on *Trichoderma reesei* cellulases has been known since 1975, when Highley (1975) showed that simple phenols such as vanillin, caffeic acid, and catechol lead to inhibition of cellulase activities. Ximenes et al. (2011) observed that β-glucosidase from *Aspergillus niger* was the most resistant enzyme to the inhibition and deactivation by phenolic compounds, requiring about five- and tenfold higher concentrations, respectively, for the same levels of inhibition or deactivation when compared with enzymes from *Trichoderma reesei*.

Regarding the type of phenolics, tannic, gallic, hydroxycinnamic, and 4-hydroxybenzoic acids, as well as vanillin and syringaldehyde (see Chap. 4), reduced the cellulase activities, including the β-glucosidase activity. Tannic acid had the most severe effect, causing both deactivation and reversible loss (inhibition) on all activities of enzymes tested (Ximenes et al. 2011). Furthermore, Tejirian and Xu (2011) found that tannic acid as low as 1 mM caused significant inhibition with approximately 70–80% decrease in cellulose hydrolysis and initial hydrolysis rate.

Kim et al. (2011, 2013) also evaluated the effect of soluble inhibitors on cellulase activity and found that both phenolic compounds and xylo-oligosaccharides were the most important factors, leading to a decrease in cellulase activity. They showed that a total phenolic compound concentration as low as 1.3 g/L strongly inhibited cellulase by the precipitation and deactivation of β-glucosidase. Xylose (21 g/L) resulted in 10% lower glucose yield and 10% lower initial hydrolysis rate compared to the control, and xylo-oligomers were stronger inhibitors than xylose. At a concentration of 8 g/L, the initial rate was approximately 40% lower, and final glucose yield was 20% lower than the control. Enzymatic hydrolysis of the xylo-oligosaccharides to xylose (21 g/L) resulted in a moderate relief of inhibition, improving hydrolysis yield by 10%. To overcome the negative effects of residual hemicellulose on enzymatic hydrolysis, other authors have used hemicellulases to boost the cellulose saccharification through (1) removal of the remaining hemicellulose and the enlarged contact area between the cellulose and the enzyme and/or (2) conversion of xylan and xylo-oligomers to the less inhibitory xylose (Romaní et al. 2014). Although sugar oligomers are cellulase inhibitors, phenolic compounds exert the dominant effect for the process of relevant conditions, unless they are removed or transformed into non-inhibitory compounds.

Michelin et al. (2016) have shown that both cellulases and hemicellulases were inhibited and/or deactivated by phenolics obtained from pretreatment of sugarcane bagasse by liquid hot water. They also highlighted that the removal or mitigation of inhibitor effects must be considered in order to improve the efficiency of cellulose hydrolysis and particularly to avoid the cascade of inhibitory effects that result when β-glucosidase or β-xylosidase activities are deactivated. Kim et al. (2011) improved the final sugar yield by 20% through the removal of phenolic compounds. Inhibition by xylose and sugar oligomers is reversible, meaning that inhibition can be reversed if sugar oligomers are removed, for example, through hydrolysis and pentose fermentation. Kumar and Wyman (2009) also showed that xylo-oligomers inhibited cellulase action and that both glucose and xylose release could be significantly enhanced by supplementation with β-xylosidase and xylanase. Other authors have reported the inhibitory effect of hemicelluloses on cellulase activity, particularly in the form of oligomers that presents a more inhibitory effect than xylan and xylose (Qing et al. 2010; Qing and Wyman 2011).

On the other hand, some phenolic compounds are known to inhibit cellulase by irreversible binding and precipitation (Kim et al. 2011; Ximenes et al. 2011). Thus, mitigating phenolic-induced inhibition/deactivation would be critical for the development of cost-efficient cellulose hydrolysis.

A number of physical and chemical methods have been developed for the removal of these inhibitory compounds including dilution, solvent extraction, precipitation, adsorption (e.g., activated charcoal), and the use of ion-exchange resins (Larsson et al. 1999; Mussatto and Roberto 2004; Carter et al. 2011). However, these detoxification methods have some drawbacks such as high cost and generation of additional waste streams. An alternative detoxification method is the use of microorganisms or their enzymes such as laccases, to remove phenolic compounds. This approach can offer improvements compared to physical and chemical methods since, in addition to other advantages (e.g., mild condition, low cost, high degradation efficiency, low energy consumption), little waste is generated (Parawira and Tekere 2011; Lee et al. 2012; Cao et al. 2013; He et al. 2016). For example, Lee et al. (2012) removed phenolic compounds from biomass slurries using a novel laccase (YlLac) from yeast *Yarrowia lipolytica*, which alleviated the cellulase inhibition and subsequently increased saccharification yield of rice straw. This detoxification method has proved to be a promising approach for improving the efficiency of biorefineries.

5.6 Directions for Enzyme Kinetic Analysis

The current research on substrate-related depolymerase activities must deal with the gap of knowledge among substrate structure, enzyme structure analysis, and their relationships to the enzyme kinetics, such as binding constants, processivity, and catalytic efficiency. The state of the art of sugarcane analysis and protein structure was heavily reviewed and is under the scope of these chapters (Driemeier et al. 2011; Yang et al. 2011; Foston and Ragauskas 2012; De Souza et al. 2013). Currently, the works of De Souza et al. (2013) and Driemeier et al. (2011) form a methodological scaffold and reference work for the study of sugarcane biomass. Here we present directions for a deep understanding of enzyme catalysis to couple it to structural studies, such as the origin and structural related controls of reaction rates, binding order, and enzyme efficiency.

5.6.1 Characterization of Enzyme Mechanisms Involved in Degradation Processes

Integrated enzyme studies related to reaction kinetics and protein dynamics are far from a comprehension of enzyme action linked to structural properties of sugarcane bagasse to better efficacy during depolymerization processes. Among these enzyme properties, the enzyme efficiency, endurance, processivity, distributivity, and dig capacity are the most valuable information to transpose the academic knowledge to the industrial sector. However, the lack of a statistical method capable of overcoming the gap among substrate structure, enzyme properties, and process yield imposes an obstacle for substrate-enzyme-yield integration. Thus, the Methods in

Enzymology remains the mandatory source of information to develop new methods to study enzymes related to depolymerization of polysaccharides in sugarcane feedstock.

Enzyme efficiency and endurance are among the major goals to be achieved for utilization in the industrial environment. The investigation of enzyme efficiency and usability is related to its kinetic, chemical, and dynamic mechanisms, a demanding task for depolymerizing enzymes. The initial-rate kinetics is the first tool for analyzing enzyme catalysis (Segel 1993). This step determines catalytic rates and constants, substrate specificity and side reactions, product inhibition and competitive inhibition—both used to analyze substrate binding order—determination of pKs of catalytic groups by pH kinetics, and analysis of the involvement of amino acid residues during catalysis by site-direct mutagenesis.

The chemical studies consider the reaction stereochemistry (NMR, X-ray crystallography); detection of tightly bound coenzymes (NMR, MS, X-ray crystallography); and metal ions (X-ray crystallography, NMR, ICP-MS) necessary to develop a specific enzyme reaction determining all organic and inorganic components necessary to develop higher reaction rates. The detection of covalent intermediates by rapid kinetic strategies and spectroscopic or MS analysis reveals the reaction course used by an enzyme, which can be integrated with its structural analysis. In turn, the identification of active-site residues involved in the binding and reaction course by affinity labelling provides information for enzyme engineering. Powerful mechanistic tools include the time-resolved Laue X-ray crystallography and video absorption spectroscopy (Purich 2010, see Chap. 6), which are useful for determining structure-function relationships in enzymes and between enzymes and substrates.

Isotope kinetics is used to investigate substrate binding order, which is extremely important to analyze and coordinate enzyme reaction events at the catalytic site. For these analyses, an enzymologist employs partial exchange reaction analysis and substrate exchange at equilibrium. Isotope kinetics is also used to study reaction "stickiness" by isotope trapping and partition kinetics, but also to identify reaction intermediates using positional isotope exchange and kinetic isotope effects (Segel 1993; Purich 2010). The use of a liquid scintillation counter and chromatographic instruments coupled to mass spectrometers or radiation counters must be employed in isotope kinetic studies.

Fast reaction kinetics is used to discriminate partial kinetic constants and reaction order to deeply understand the reaction events at the catalytic site, determining the time-dependent bottleneck events during the reaction course of a processing enzyme. The design of experiments for enzymes acting on polymeric substrates is challenging. These studies employ special equipment, and encompass the continuous, stopped-flow, and mix/quenching techniques, the temperature-jump and pressure-jump relaxation techniques, and the cryoenzymology. These techniques are used to obtain data about reaction steps on a shorter timescale inferior to 1 s. While stopped-flow devices are useful at 5–10 ms, faster reactions require the use of relaxation techniques, cryotemperatures, or even NMR and photon-activated experiments.

Depolymerases acting by processive or distributive mechanisms generally possess "hidden" nonproductive interactions in steady-state treatments or shifted binding behavior at high substrate concentration, a problem for process efficiency studies. Many exopolymerases were listed as highly processive (Purich 2010), remaining attached to their polymeric substrate during several catalytic rounds. The higher the processivity of the enzyme, the greater is its permanence in the substrate holding productive interactions, yielding monosaccharides for fermentation. Distributive mechanisms are characterized by the release of enzyme molecules after each catalytic cycle. In turn, these enzymes are capable of fragmenting the substrate, a typical behavior of endoglucanases. Monitoring these activities are conclusive kinetic studies to identify an exopoligalacturonase or endopoligalacturonase and its catalytic efficiencies. Protocols to determine processivity were presented by McClure and Chow (1980). The basic analytic tool measures discrete product lengths, which are studied using the Kuhn distribution law. Random scission kinetics of some endo-depolymerases (Purich 2010) can be analyzed using a modification of the protocol of Thomas (1956) analyzing the DP of cleavage products by means of GPC or MS. Purich (2010) presented an initial source of mathematical treatments to analyze nonproductive interactions and shifted binding behavior. In turn, the dig of cell walls, an important mechanism to access the deep layers of the cell wall by depolymerases, can be image measured using ATM, SEM, or electron tomography. The current lack of these knowledge is the main reason for failures in operational plants for bioethanol production.

5.7 Concluding Remarks

Sugarcane has emerged as a leading culture in Brazil for the development of new technologies for biofuels in order to ensure the energy demand in a sustainable manner (see Chap. 10 for further details).

Currently, a second-generation ethanol production plant is operating in Brazil, the Bioflex 1 from GranBio with a production capacity of 82 million liters of ethanol per year (Elabora Consultoria 2014). Internationally, the major players in cellulosic ethanol are INEOS Bio, KiOR, Novozymes, Petrobrás, Dupont, Raysen-CTC, and POET. Logen Corporation and Beta Renewables are responsible for pretreatment of biomass in the Brazilian industries of cellulosic ethanol. Their income with current substrate variability or high production costs will be of fundamental importance to guide research on the development of processes of pretreatments.

Besides these aspects, the relevance of the sugarcane bioenergy for the social and economic development of Central America, South America, and sub-Saharan Africa is immense (see Chap. 11). These regions are extremely favorable for the sustainable expansion of bioenergy, since there is no land available for the substantial production of bioenergy and food in other regions. The proper land management can result in production capacity of more than 24 EJ of liquid biofuels without compromising food production (Souza et al. 2015). Thus, the assurance of the energy security of our century includes the development of emerging or underdeveloped countries with potential impact on the eradication of poverty.

Acknowledgements We thank Fundação de Amparo à Pesquisa do Estado de São Paulo (FAPESP, process 2010/52322-3) and Conselho de Desenvolvimento Científico e Tecnológico (CNPq, process 563260-6). This project is also part of National Institute of Science and Technology of the Bioethanol (FAPESP, process 2008/57908-6). Dr. M. L. T. M. Polizeli has a Fellowship of Research Productivity of CNPq. We thank Abilio Borghi for the technical assistance with the English language.

References

Alvira P, Tomás-Pejó E, Ballesteros M, Negro MJ (2010) Pretreatment technologies for an efficient bioethanol production process based on enzymatic hydrolysis: a review. Bioresour Technol 101:4851–4861

ANFAVEA (n.d.) Associação Nacional dos Fabricantes de Veículos automotores, Brasil. Dados estatísticos disponíveis em http://www.anfavea.com.br

Berlin A, Balakshin M, Gilkes N, Kadla J, Maximenko V, Kubo S, Saddler J (2006) Inhibition of cellulase, xylanase and beta glucosidase activities by softwood lignin preparations. J Biotechnol 125:198–209

BNDES (2015) O Déficit de Produção de etanol no Brasil entre 2012 e 2015: determinantes, consequências e sugestões de política. Milanez AY, Nyko D, Garcia JLF, Reis BLSFS. http://www.bndes.gov.br/bibliotecadigital. Accessed Mar 2015

Bon EPS, Gírio F, Pereira N Jr (2008) Enzimas na produção de etanol. In: Bon EPS, Ferrara MA, Corvo ML, Vermelho AB, Paiva CLA, Alencastro RB, Coelho RRR (eds) Enzimas em Biotecnologia Produção, Aplicações e Mercado. Editora Interciência, Rio de Janeiro (Chapter 10)

Borin GP, Sanchez CC, De Souza AP, Santana ES, Souza AT, Leme AFP, Squina FM, Buckeridge MS, Goldman GH, Oliveira JVC (2015) Comparative secretome analysis of *Trichoderma reesei* and *Aspergillus niger* during growth on sugarcane biomass. PLoS One 10(6):e0129275

Brodeur G, Yau E, Badal K, Collier J, Ramachandran KB, Ramakrishnan S (2011) Chemical and physicochemical pretreatment of lignocellulosic biomass: a review. Enzyme Res 2011:1–17

Buckeridge MS (2010) Seed cell wall storage polysaccharides: models to understand cell wall biosynthesis and degradation. Plant Physiol 154:1017–1023

Buckeridge MS, De Souza AP (2014) Breaking the "Glycomic Code" of cell wall polysaccharides may improve second-generation bioenergy production from biomass. Bioenergy Res 7: 1065–1073

Buckeridge MS, Santos HP, Tiné MAS (2000) Mobilisation of storage cell wall polysaccharides in seeds. Plant Physiol Biochem 38:141–156

Buckeridge MS, Santos WD, Tiné MS, De Souza AP (2015) The cell wall architecture of sugarcane and its implications to cell wall recalcitrance. In: Lam E, Carrer H, Silva JA (eds) Compendium of bioenergy plants: sugarcane. CRC Press, Boca Raton, p 125

Cannella D, Hsieh CW, Felby C, Jørgensen H (2012) Production and effect of aldonic acids during enzymatic hydrolysis of lignocellulose at high dry matter content. Biotechnol Biofuels 5(1):26

Cantarel BL, Coutinho PM, Rancurel C, Bernard T, Lombard V, Henrissat B (2009) The Carbohydrate-Active EnZymes database (CAZy): an expert resource for glycogenomics. Nucleic Acids Res 37:233–238

Cao G, Ximenes E, Nichols NN, Zhang L, Ladisch M (2013) Biological abatement of cellulase inhibitors. Bioresour Technol 146:604–610

Carter B, Squillace P, Gilcrease PC, Menkhaus TJ (2011) Detoxification of a lignocellulosic biomass slurry by soluble polyelectrolyte adsorption for improved fermentation efficiency. Biotechnol Bioeng 108:2053–2060

Castro MD, Pereira N Jr (2010) Produção, propriedades e aplicação de celulases na hidrólise de resíduos agroindustriais. Quím Nova 33:181–188

Caufrier F, Martinou A, Dupont C, Bouriotis V (2003) Carbohydrate esterase family 4 enzymes: substrate specificity. Carbohydr Res 338(7):687–692

Crepin VF, Fauld CB, Connerton IF (2004) Functional classification of the microbial feruloyl esterases. Appl Microbiol Biotechnol 63(6):647–652

Damasio ARL, Silva TM, Almeida FBDR, Squina FM, Ribeiro DA, Leme AFP, Segato F, Prade RA, Jorge JA, Terenzi HF, Polizeli MLTM (2011) Heterologous expression of an *Aspergillus niveus* xylanase GH11 in *Aspergillus nidulans* and its characterization and application. Process Biochem 46:1236–1242

Damasio ARL, Pessela BC, Segato F, Prade RA, Guisan JM, Polizeli MLTM (2012a) Improvement of fungal arabinofuranosidase thermal stability by reversible immobilization. Process Biochem 47(12):2411–2417

Damasio ARL, Polizeli MLTM, Guisan JM, Prade RA, Pessela BC, Mateo C, Segato F (2012b) Immobilization of a recombinant endo-1,5-arabinanase secreted by *Aspergillus nidulans* strain A773. J Mol Catal B: Enzym 77:39–45

Damasio ARL, Ribeiro LFC, Ribeiro LF, Furtado GP, Segato F, Almeida FB, Crivellari AC, Buckeridge MS, Souza TA, Murakami MT, Ward RJ, Prade RA, Polizeli MLTM (2012c) Functional characterization and oligomerization of a recombinant xyloglucan-specific endo-β-1,4-glucanase (GH12) from *Aspergillus niveus*. Biochim Biophys Acta 1824:461–467

Damasio ARL, Pessela BC, Silva TM, Guimaraes LH, Jorge JA, Guisan JM, Polizeli MLTM (2013) Co-immobilization of fungal endo-xylanase and α-L-arabinofuranosidase in glyoxyl agarose for improved hydrolysis of arabinoxylan. J Biochem 154:275–280

De Lucas RC, Cereia M (2016) Mycology: current and future developments. In: Chapter 1. Large scale production of cellulases for biomass degradation. Bentham Science, Sharjah, vol 1, pp 316–328

De Paula RG, Antoniêto ACC, Silva RN (2016) Mycology: current and future developments. In: Chapter 1. The panorama for biofuels biotechnology: present and future. Bentham Science, Sharjah, vol 1, pp 3–22

De Souza AP, Leite DCC, Pattathil S, Hahn MG, Buckeridge MS (2013) Composition and structure of sugarcane cell wall polysaccharides: implications for second-generation bioethanol production. Bioenergy Res 6:564–579

De Souza AP, Grandis A, Leite DCC, Buckeridge MS (2014) Sugarcane as a bioenergy source: history, performance, and perspectives for second-generation bioethanol. Bioenergy Res 7: 24–35

De Vries RP, Visser J (2001) Aspergillus enzymes involved in degradation of plant cell wall polysaccharides. Microbiol Mol Biol Rev 65(4):497–522

Dodd D, Cann IK (2009) Enzymatic deconstruction of xylan for biofuel production. Glob Change Biol Bioenergy 1(1):2–17

Donohoe BS, Decker SR, Tucker MP, Himmel ME, Vinzant TB (2008) Visualizing lignin coalescence and migration trough maize cell walls following thermochemical pretreatment. Biotechnol Bioeng 101:913–925

Driemeier C, Pimenta MTB, Rocha GJM, Oliveira MM, Mello DB, Maziero P, Gonçalvess AR (2011) Evolution of cellulose crystals during prehydrolysis and soda delignification of sugarcane lignocellulose. Cellulose 18:1509–1519

DuPont (2013) Acellerase 1500®: cellulase enzyme complex for lignocellulosic biomass hydrolysis. http://accellerase.dupont.com/fileadmin/user_upload/live/accellerase/documents/DUP-00413_ProdSheet_1500_web.pdf

Elabora Consultoria (2014) Renewable fuels and chemicals. Elabora Editora, Sao Paulo

Eriksson T, Karlsson J, Tjerneld F (2002) A model explaining declining rate in hydrolysis of lignocellulose substrates with cellobiohydrolase I (Cel7A) and endoglucanase I (Cel7B) of *Trichoderma reesei*. Appl Biochem Biotechnol 101:41–60

Foston M, Ragauskas AJ (2012) Biomass characterization: recent progress in understanding biomass recalcitrance. Ind Biotechnol 8(4):191–208

Furtado GP, Santos CR, Cordeiro RL, Ribeiro LF, de Moraes LAB, Damasio ARL, Polizeli MLTM, Lourenzoni MR, Murakami MT, Ward RJ (2015) Enhanced xyloglucan-specific endo-β-1,4-glucanase efficiency in an engineered CBM44-XegA chimera. Appl Microbiol Biotechnol 99(12):5095–5107

Gilkes NR, Henrissat B, Kilburn DG, Miller RC Jr, Warren RAJ (1991) Domains in microbial beta-1-4-glycanases: sequence conservation, function, and enzyme families. Microbiol Rev 55:303–315

Gírio FM, Fonseca C, Carvalheiro F, Duarte LC, Marques S, Bogel-Łukasik R (2010) Hemicelluloses for fuel ethanol: a review. Bioresour Technol 101(13):4775–4800

He H, Zhang J, Bao J (2016) Acceleration of biodetoxification on dilute acid pretreated lignocellulose feedstock by aeration and the consequente ethanol fermentation evaluation. Biotechnol Biofuels 9:19

Hemsworth GR, Taylor EJ, Kim RQ, Gregory RC, Lewis SJ, Turkenburg JP, Parkin A, Davies GJ, Walton PH (2013) The copper active site of CBM33 polysaccharide oxygenases. J Am Chem Soc 135(16):6069–6077

Hemsworth GR, Henrissat B, Davies GJ, Walton PH (2014) Discovery and characterization of a new family of lytic polysaccharide monooxygenases. Nat Chem Biol 10(2):122–126

Henrissat B, Davies GJ (2000) Glycoside hydrolases and glycosyltransferases. Families, modules, and implications for genomics. Plant Physiol 124:1515–1519

Highley TL (1975) Inhibition of cellulases of wood decay fungi. US Forest Service Research Paper FPL 247, 1–8, 1975. http://accellerase.dupont.com/fileadmin/user_upload/live/accellerase/documents/DUP-00413_ProdSheet_1500_web.pdf

Hu J, Arantes V, Pribowo A, Saddler JN (2013) The synergistic action of accessory enzymes enhances the hydrolytic potential of a "cellulase mixture" but is highly substrate specific. Biotechnol Biofuels 6(1):112

Ishizawa CI, Jeoh T, Adney WS, Himmel ME, Johnson DK, Davis MF (2009) Can delignification decrease cellulose digestibility in acid pretreated corn stover? Cellulose 16:677–686

Jing X, Xhang X, Bao J (2009) Inhibition performance of lignocellulose degradation products on industrial cellulase enzymes during cellulose hydrolysis. Appl Biochem Biotechnol 159:696–707

Kallioinen A (2014) Development of pretreatment technology and enzymatic hydrolysis for biorefineries. VTT Technical Research Centre of Finland

Kim Y, Ximenes E, Mosier NS, Ladisch MR (2011) Soluble inhibitors/deactivators of cellulase enzymes from lignocellulosic biomass. Enzyme Microb Technol 48:408–415

Kim Y, Kreke T, Hendrickson R, Parenti J, Ladisch MR (2013) Fractionation of cellulase and fermentation inhibitors from steam pretreated mixed hardwood. Bioresour Technol 135:30–38

Kjaergaard CH, Qayyum MF, Wong SD, Xu F, Hemsworth GR, Walton DJ, Young NA, Davies GJ, Walton PH, Johansen KS, Hodgson KO, Hedman B, Solomon EI (2014) Spectroscopic and computational insight into the activation of O_2 by the mononuclear Cu center in polysaccharide monooxygenases. Proc Natl Acad Sci U S A 111(24):8797–8802

Kumar R, Wyman CE (2009) Effect of xylanase supplementation of cellulase on digestion of corn stover solids prepared by leading pretreatment technologies. Bioresour Technol 100:4203–4213

Kumar P, Barrett DM, Delwiche MJ, Stroeve P (2009) Methods for pretreatment of lignocellulosic biomass for efficient hydrolysis and biofuel production. Ind Eng Chem Res 48(8):3713–3729

Kurakabe M, Shinjii O, Komaki T (1997) Transxylosilation of β-xylosidase from *Aspergillus awamori* K4. Biosci Biotechnol Biochem 6112:2010–2014

Larsson S, Reimann A, Nilvebrant NO, Jönsson LJ (1999) Comparison of different methods for the detoxification of lignocellulose hydrolyzates of spruce. Appl Biochem Biotechnol 77:91–103

Lee K-M, Kalyani D, Tiwari MK, Kim T-S, Dhiman SS, Lee J-K, Kim I-W (2012) Enhanced enzymatic hydrolysis of rice straw by removal of phenolic compounds using a novel laccase from yeast *Yarrowia lipolytica*. Bioresour Technol 123:636–645

Li X, Chomvong K, Yu VY, Liang JM, Lin Y, Cate JH (2015) Cellobionic acid utilization: from *Neurospora crassa* to *Saccharomyces cerevisiae*. Biotechnol Biofuels 8:120

Lo Leggio L, Simmons TJ, Poulsen JC, Frandsen KE, Hemsworth GR, Stringer MA, von Freiesleben P, Tovborg M, Johansen KS, De Maria L, Harris PV, Soong CL, Dupree P, Tryfona T, Lenfant N, Henrissat B, Davies GJ, Walton PH (2015) Structure and boosting activity of a starch-degrading lytic polysaccharide monooxygenase. Nat Commun 6:5961

Lynd LR, Laser MS, Brandsby D, Dale BE, Davison B, Hamilton R, Himmel M, Keller M, McMillan JD, Sheehan J, Wyman CE (2008) How biotech can transform biofuels. Nat Biotechnol 26:169–172

McClure WR, Chow Y (1980) The kinetics and processivity of nucleic acid polymerases. Methods Enzymol 64:277–297

McMillan JD, Jennings EW, Mohagheghi A, Zuccarello M (2011) Comparative performance of precommercial cellulases hydrolyzing pretreated corn stover. Biotechnol Biofuels 4:29

Menon V, Rao M (2012) Trends in bioconversion of lignocellulose: biofuels, platform chemicals & biorefinery concept. Prog Energy Comb Sci 38:522–550

Michelin M, Polizeli MLTM, Ruzene DS, Silva DP, Teixeira JA (2014) Application of lignocellulosic residues in the production of cellulases and hemicellulases from fungi. In: Polizeli MLTM, Rai M (eds) Fungal enzymes. CRC Press, Boca Raton, pp 31–64

Michelin M, Ximenes E, Polizeli MLTM, Ladisch MR (2016) Effect of phenolic compounds from pretreated sugarcane bagasse on cellulolytic and hemicellulolytic activities. Bioresour Technol 199:275–278

Miranda, E.E. (2010) Questões Ambientais (Impactos Locais e Globais) e energéticas: A expansão da Cana de Açúcar e a ocupação das terras no estado de São Paulo.In: Cortez LAB (ed) Bioetanol de Cana-de-açúcar—P&D para produtividade e sustentabilidade. Editora Blucher (Chapter 6)

Mussatto SI, Roberto IC (2004) Alternatives for detoxification of diluted-acid lignocellulosic hydrolyzates for use in fermentative processes: a review. Bioresour Technol 93:1–10

Nacke H, Engelhaupt M, Brady S, Fischer C, Tautzt J, Daniel R (2012) Identification and characterization of novel cellulolytic and hemicellulolytic genes and enzymes derived from German grassland soil metagenomes. Biotechnol Lett 34(4):663–675

Nakagame S, Chandra RP, Saddler JN (2010) The effect of isolated lignins, obtained from a range of pretreated lignocellulosic substrates, on enzymatic hydrolysis. Biotechnol Bioeng 105(5): 871–879

Novozymes (2012) Cellulosic ethanol: novozymes cellic Ctec3. http://www.bioenergy.novozymes.com/en/cellulosic-ethanol/Cellic-HTec3/Documents/CE_APP_Cellic_Ctec3.pdf

Pakarinen A, Haven MO, Djajadi DT, Várnai A, Puranen T, Viikari L (2014) Cellulases without carbohydrate-binding modules in high consistency ethanol production process. Biotechnol Biofuels 7(1):27

Parawira W, Tekere M (2011) Biotechnological strategies to overcome inhibitors in lignocellulose hydrolysates for ethanol production: review. Crit Rev Biotechnol 31:20–31

Payne CM, Knott BC, Mayes HB, Hansson H, Himmel ME, Sandgren M, Ståhlberg J, Beckham GT (2015) Fungal cellulases. Chem Rev 115(3):1308–1448

Polizeli MLTM, Rizzatti ACS, Monti R, Terenzi HF, Jorge JA, Amorim DS (2005) Xylanases from fungi: properties and industrial applications. Appl Microbiol Biotechnol 67:577–591

Pollet A, Delcour JA, Courtin CM (2010) Structural determinants of the substrate specificities of xylanases from different glycoside hydrolase families. Crit Rev Biotechnol 30(3):176–191

Purich DL (2010) Enzyme kinetics: catalysis & control. Academic Press, London

Qing Q, Wyman CE (2011) Supplementation with xylanase and β-xylosidase to reduce xylo-oligomer and xylan inhibition of enzymatic hydrolysis of cellulose and pretreated corn stover. Biotechnol Biofuels 4:18

Qing Q, Yang B, Wyman CE (2010) Xylooligomers are strong inhibitors of cellulose hydrolysis by enzymes. Bioresour Technol 101:9624–9630

Rahikainen J, Mikander S, Marjamaa K, Tamminen T, Lappas A, Viikari L, Kruus K (2011) Inhibition of enzymatic hydrolysis by residual lignins from softwood—study of enzyme binding and inactivation on lignin-rich surface. Biotechnol Bioeng 108(12):2823–2834

Rani V, Mohanram S, Tiwari R, Nain L, Arora A (2014) Beta-glucosidase: key enzyme in determining efficiency of cellulase and biomass hydrolysis. J Bioproces Biotech 5(1):8

Ribeiro LFC, Lucas RC, Vitcosque GL, Ribeiro LF, Ward RJ, Rubio MV, Damasio ARL, Gregory RC, Walton PH, Jorge JA, Prade RA, Buckeridge MS, Polizeli MLTM (2014) A novel thermostable xylanase GH10 from *Malbranchea pulchella* expressed in *Aspergillus nidulans* with potential applications in biotechnology. Biotechnol Biofuels 7:115–126

Rocha GJM, Gonçalves AR, Oliveira BR, Olivares EG, Rossella CEV (2011) Steam explosion pretreatment reproduction and alkaline delignification reactions performed on a pilot scale with sugarcane bagasse for bioethanol production. Ind Crop Prod 35:274–279

Rollin JA, Zhu Z, Sathitsuksanoh N, Zhang YHP (2011) Increasing cellulose accessibility is more important than removing lignin: a comparison of cellulose solvent-based lignocellulose fractionation and soaking in aqueous ammonia. Biotechnol Bioeng 108:22–30

Romaní A, Ruiz HA, Pereira FB, Teixeira JA, Domingues L (2014) Integrated approach for effective bioethanol production using whole slurry from autohydrolyzed *Eucalyptus globulus* wood at high-solid loadings. Fuel 135:482–491

Santos FA, Queiróz JH, Colodette JL, Fernandes SA, Guimarães VM, Rezende ST (2012) Potencial da palha de cana-de-açúcar para produção de etanol. Quim Nova 35:1004–1010

Segato F, Damasio AR, Gonçalves TA, Murakami MT, Squina FM, Polizeli MLTM, Mort AJ, Prade RA (2012) Two structurally discrete GH7-cellobiohydrolases compete for the same cellulosic substrate fiber. Biotechnol Biofuels 5:21

Segato F, Damasio ARL, de Lucas RC, Squina FM, Prade RA (2014) Genomics review of holocellulose deconstruction by Aspergilli. Microbiol Mol Biol Rev 78:588–613

Segel IH (1993) Enzyme kinetics: behavior and analysis of rapid equilibrium and steady-state enzyme systems. Wiley, New York

Soccol RC, Vandenberghe LPS, Medeiros ABP, Karp SG, Buckeridge M, Ramos LP, Pitarelo AP, Ferreira-Leitão V, Gottschalk LMF, Ferrara MA, Bon EPS, Moraes LMP, Araújo JA, Torres FAG (2010) Bioethanol from lignocelluloses: status and perspectives in Brazil. Bioresour Technol 101:4820–4825

Souza, G. M., Victoria, R. L., Joly, C. A., Verdade, L. M. (eds) (2015) Bioenergy & sustainability: bridging the gaps. Scientific Committee on Problems of the Environment (SCOPE)

Tavares EQP, Buckeridge MS (2015) Do cell walls have a code? Plant Sci 241:286–294

Tejirian A, Xu F (2011) Inhibition of enzymatic cellulolysis by phenolic compounds. Enzyme Microb Technol 48:239–247

Tenkanen M, Siika-Aho M (2000) An alpha-glucuronidase of *Schizophyllum commune* acting on polymeric xylan. J Biotechnol 78(2):149–161

Thomas CA Jr (1956) The enzymatic degradation of desoxyribose nucleic acid. J Am Chem Soc 78:1861–1868

UNICA (2015) from Brazil's National Institute for Space Research. http://www.unica.com.br/unicadata. Accessed 4 June 2016

Vitcosque GL, Ribeiro LF, de Lucas RC, da Silva TM, Ribeiro LF, Damasio ARL, Farinas CS, Gonçalves AZ, Segato F, Buckeridge MS, Jorge JA, Polizeli MLTM (2016) The functional properties of a xyloglucanase (GH12) of Aspergillus terreus expressed in Aspergillus nidulans may increase performance of biomass degradation. Appl Microbiol Technol 100(21):9133–9144. doi:10.1007/s00253-016-7589-2

Vuong TV, Wilson DB (2010) Glycoside hydrolases: catalytic base/nucleophile diversity. Biotechnol Bioeng 107(2):195–205

Ximenes E, Kim Y, Mosier N, Dien B, Ladisch M (2010) Inhibition of cellulases by phenols. Enzyme Microb Technol 46:170–176

Ximenes E, Kim Y, Mosier N, Dien B, Ladisch M (2011) Deactivation of cellulases by phenols. Enzyme Microb Technol 48:54–60

Yang B, Willies DM, Wyman CE (2006) Changes in the enzymatic hydrolysis rate of avicel cellulose with conversion. Biotechnol Bioeng 94:1122–1128

Yang B, Dai Z, Ding S, Wyman CE (2011) Enzymatic hydrolysis of cellulosic biomass. Biofuels 2(4):421–450

Zhang YP, Lynd LR (2004) Toward an aggregated understanding of enzymatic hydrolysis of cellulose: noncomplexed cellulase systems. Biotechnol Bioeng 88:797–824

Zhang PYH, Himmel ME, Mielenz JR (2006) Outlook for cellulase improvement: screening and selection strategies. Biotechnol Adv 24(5):452–481

Chapter 6
The Role of Fungal Transcriptome Analysis and Side-Chain Hydrolyzing Enzymes in Sugarcane Bagasse Breakdown

Leonora Rios de Souza Moreira, Camila Louly Corrêa, Helder Andrey Rocha Gomes, Glaucia Emy Okida Midorikawa, Robert Neil Gerard Miller, and Edivaldo Ximenes Ferreira Filho

Abstract Lignocellulose represents one of the main sources of renewable energy for biofuel production. Within this context, sugarcane bagasse, which is a discarded by-product of sugarcane processing, is a rich source of lignin, hemicellulose, and cellulose. Since sugarcane bagasse is a complex lignocellulosic structure, its complete enzymatic hydrolysis requires the action of an enzyme system of main- and side-chain enzymes. The xylan component in sugarcane bagasse is arabinosylated at positions C-2 and/or C-3 and esterified with ferulic acid at the C-5 position. Accessory enzymes such as α-arabinofuranosidase and feruloyl esterase are responsible for enhanced accessibility and enzymatic hydrolysis of xylan. Under industrial conditions, numerous filamentous fungi can secrete enzymes or enzyme complexes belonging to the glycosyl hydrolases, carbohydrate esterases, and polysaccharide lyases for the hydrolysis of cellulose, hemicellulose, and lignin. Next-generation sequencing approaches for whole-genome and transcriptome analyses enable the identification of novel genes and isoforms in fungi encoding hydrolytic enzymes, thus increasing our understanding of mechanisms involved in gene expression induction in the presence of plant cell wall or its metabolic products. The downstream application will accelerate strain improvement for increased enzymatic efficiencies and hydrolytic enzyme cocktail development, contributing to reducing cellulosic bioethanol production costs.

Keywords Arabinofuranosidase • Feruloyl-esterase • Side-chain enzymes • Transcriptome • Sugarcane bagasse

---

[†]These authors contributed equally to this work.

L.R. de Souza Moreira • C.L. Corrêa • H.A.R. Gomes
G.E.O. Midorikawa • R.N.G. Miller • E.X.F. Filho (✉)
Department of Cellular Biology, University of Brasília, Brasília, DF CEP 70910-900, Brazil
e-mail: eximenes@unb.br

M.S. Buckeridge, A.P. De Souza (eds.), *Advances of Basic Science for Second Generation Bioethanol from Sugarcane*, DOI 10.1007/978-3-319-49826-3_6

6.1 Introduction

In times of growing concern over the limitation of fossil fuels, the utilization of renewable resources such as ethanol has become of great interest. The increased demand for energy, geopolitical factors, and environmental pressures have alarmed the scientific community, economists, and governments to evaluate the potential of cellulosic ethanol as a sustainable, economic, and eco-friendly alternative to gasoline. One of the main sources of renewable energy for biofuels is the conversion of plant-derived carbohydrates into bioethanol. Brazil plays an important role today in meeting the world's demands for bioethanol, with sugarcane-derived ethanol accounting for approximately 30% of the world's production. Sugarcane bagasse is an excellent raw material for second-generation ethanol production in Brazil and other countries such as India and China, where this biomass residue is generated in large amounts every year (Goldemberg 2007; Chandel et al. 2012, 2013; de Souza et al. 2014).

Sugarcane bagasse is a fibrous residue of cane stalks left over after the crushing and extraction of the juice from the sugarcane. This residue is produced in large quantities by sugar and alcohol industries. In general, 1 ton of sugarcane generates 280 kg of bagasse (Sun et al. 2004b; Pandey et al. 2000).

The stockpiled bagasse is of low economic value and constitutes an environmental problem to sugar mills and surrounding districts, especially if stocked for extended periods due to the risk of spontaneous combustion occurring within the pile (de Souza et al. 2013). Chemically, about 40–50% of the dry residue of sugarcane corresponds to the glucose polymer cellulose, most of which is in a crystalline structure. Hemicelluloses, which include xylans and mannans, represent the second most abundant fraction (25–35%). The remainder is mostly lignin and lesser amounts of minerals, waxes, and other compounds (Jacobsen and Wyman 2002). Another study reports 42% cellulose, 22% lignin, 28% hemicelluloses, and 8% of cane wax and organic acid (Beukes et al. 2008). Chemical composition varies across the different plant tissues of sugarcane stalks, with greater lignin and cellulose content in epidermal tissues in comparison with nodes and internodes (Brienzo et al. 2014). Because of its low ash content (2.3%), bagasse offers advantages for application as a novel material for industries when compared with other crop residues such as rice straw, which has 13.3% ash content (Sun et al. 2004a).

Saprophytic filamentous fungi are known for their potential in decomposition of plant lignocellulosic biomass to simple fermentable sugars. A wide variety of fungi, especially those belonging to the Basidiomycete and Ascomycete classes, have been the target of investigation by the second-generation bioethanol industry. A large number and variety of secretory enzymes or enzyme complexes have been characterized in such organisms, which are applicable to hydrolysis of cellulose and hemicellulose (Couturier et al. 2012). Numerous studies have shown that GH expression in filamentous fungi usually requires induction. In order to obtain this induction, lignocellulosic materials can be used as a carbon source in the culture medium, enabling the fungal growth and the induction of cellulase and hemicellulase enzymes (Sørensen et al. 2011).

The enzymes involved in the degradation of plant polysaccharides can be divided into different glycosidase families based on the amino acid sequences structurally related to the catalytic module (CAZy database). These comprise at least 35 families of glycosyl hydrolases (GH), 3 families of carbohydrate esterases (EC), and 6 families of polysaccharide lyases (PL) (Battaglia et al. 2011). In addition to the wide variety of enzymes with different specificities that are needed to degrade components of lignocelluloses, additional proteins that contribute to the degradation of the plant cell wall, such as glycosyl hydrolase family members (GH 61) and expansions and swollenins, have also been identified (van Dyk and Pletschke 2012).

6.2 Sugarcane Bagasse Structure

Sugarcane is today the most efficient natural raw material for first-generation (1G) bioethanol production by distillation, with a much lower consumption of fossil energy during processing than that required for maize (Macedo et al. 2008). One of the by-products generated during processing is sugarcane bagasse, which is usually burned to produce steam and electricity, providing the energy needed to meet the requirements of the distillation process. Efficient cogeneration technologies for the production of 1G bioethanol still result in surplus bagasse that can be applied as a primary material for the production of 2G bioethanol or other biological based products, in addition to using it as a fuel source for electricity generation (Ensinas et al. 2007).

Many studies have been conducted to characterize the hemicellulose and lignin fractions present in the cell wall of sugarcane bagasse, and some of the data available in the literature are summarized in Table 6.1. Xylose represents the major monosaccharide present in hemicellulose, followed by arabinose. In bagasse, the hemicellulose is composed of a xylan polymer onto which other groups are bound, commonly glucuronic acid and arabinose (Lavarack et al. 2002). The most abundant hemicellulose in annual plants such as sugarcane is the arabinoxylan (Fig. 6.1), which is composed of a backbone of β-(1,4)-linked xylosyl residues, substituted with arabinoses and 4-*O*-methyl-D-glucuronic acid at C-2 and/or C-3 of the main chain (Xu et al. 2006; Peng et al. 2009). The higher arabinose content present in bagasse indicates a higher degree of branching of the xylan chains, and so higher solubility of the polymers (Sun et al. 2004a). Arabinose also plays an important role in the cross-links between hemicelluloses and lignin (Sun et al. 2011).

Degraded hemicelluloses from sugarcane bagasse also contain minor quantities of bound lignin and hydroxycinnamic acids, such as ferulic and *p*-coumaric acids (Xu et al. 2006). In the lignin fraction, ferulic acid, 3-(3-methoxy-4-hydroxyphenyl)-2-propenoic acid, and *p*-coumaric acid are the major hydroxycinnamic acids found in the cell walls (Xu et al. 2005). Ferulic acid, being one of the abundant hydroxycinnamic acids in plant cell walls, is a very attractive phenolic compound, as it can be used as an antioxidant or be transformed by microbial conversion into "natural" vanillin, an expensive flavor for the food, cosmetic, and pharmaceutical industries

Table 6.1 Monosaccharide contents in the hemicellulosic fraction from sugarcane bagasse

Method of extraction	Arabinose	Rhamnose	Xylose	Mannose	Galactose	Glucose	Reference
Water-soluble hemicellulose and lignin fraction	12.83	1.25	37.38	8.07	11.61	28.86	Sun et al. (2004a)
Alkali-soluble hemicellulose and lignin fraction	12.57	0.42	80.60	0.40	1.83	4.18	Sun et al. (2004a)
0.5 M NaOH and 0.5 M H_2O_2	10.40	0.44	78.98	0.32	1.95	7.93	Sun et al. (2004a)
Alcohol-insoluble residue hydrolyzed with H_2SO_4	2.4	N.D	33.5	N.D	0.9	59.8	De Souza et al. (2013)
Acid hydrolysates	1.80	ND	20.30	ND	0.65	36.74	Szczerbowski et al. (2014)
Dewaxed bagasse with 1 M NaOH aqueous solution at 40 °C/18 h	11.6	6.5	78	1.4	0.3	2.2	Xu et al. (2006)
Precipitated with 30% ethanol and solubilized with 3% NaOH	5.61	ND	92.98	Traces	ND	1.35	Peng et al. (2009)
Precipitated with 15% ethanol and solubilized with 1% NaOH	12.13	0.13	2.77	0.53	2.77	5.20	Peng et al. (2009)

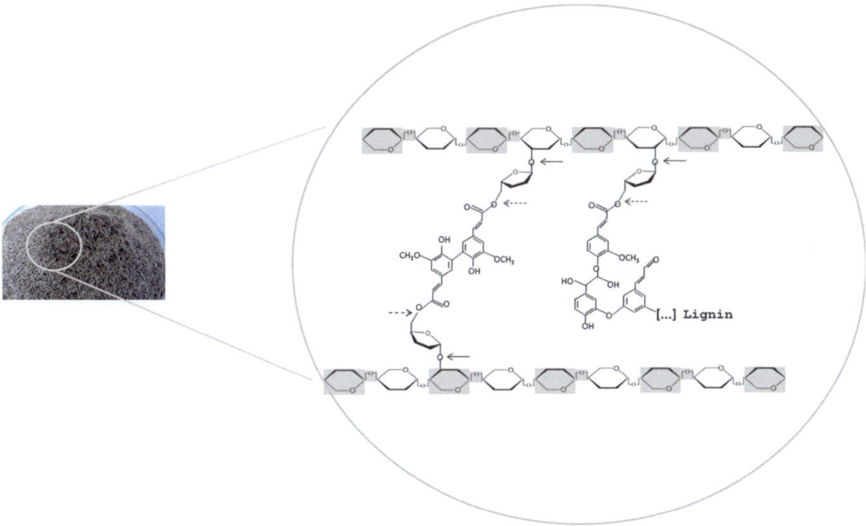

Fig. 6.1 The structure of feruloylated arabinoxylan, with diferulic bridges. Cleavage sites for α-L-arabinofuranosidase (*full arrow*) and feruloyl esterase (*dashed arrow*)

(Levasseur et al. 2005). Table 6.2 shows the content of phenols of the associated lignin in isolated hemicellulosic fractions.

Ferulic and *p*-coumaric acids are found covalently linked to polysaccharides by ester bonds, and to components of lignin by ester or ether bonds. A considerable proportion of *p*-coumaric acid is known to be esterified with lignin, while ferulic acid is extensively etherified with lignin (Xu et al. 2005; Sun et al. 2011). During the cell wall biosynthesis, ferulated arabinoxylans are extensively cross-linked by the coupling of ferulic into diferulates and by the copolymerization of ferulate and diferulates with monolignols to form polysaccharide-lignin complexes (Ralph et al. 2004). Ferulic acids, which are associated both with lignin by ether bond and with hemicelluloses by ester linkage, contribute to the cell wall cross-linking (Xu et al. 2005, for more details see Chap. 4).

Considering the biotechnological potential of lignocellulosic fungi, where applications include the food, feed, and textile industries, pulp and paper treatment, and in particular the biofuel industry, a complete understanding is needed regarding substrate-specific enzyme expression, regulatory mechanisms, and enzyme expression levels according to specific metabolic pathways. Through functional genomics or transcriptomics, it is now possible to predict the biotechnological potential of particular fungal strains according to the identification of genes and comparison of gene expression levels under different treatments. Although numerous studies have been conducted to characterize enzymes involved in the degradation of lignocellulose, understanding about transcription and regulation of genes encoding these enzymes is still incomplete. Thus, studies on the mechanisms of regulation and expression of such genes in candidate fungi are essential for the development of viable second-generation bioethanol processes.

Table 6.2 The percent of phenolic acids of the lignin fraction associated to hemicellulose from sugarcane bagasse

Method of extraction	pHBA[a]	Vanillic acid	Vanillin	Syringic acid	*p*-Coumaric acid	Ferulic acid	Reference
Water-soluble hemicellulose and lignin fraction	0.59	0.03	1.05	0.04	0.1	0.06	Sun et al. (2004a)
Alkali-soluble hemicellulose and lignin fraction	0.10	0.008	0.36	0.02	0.011	0.03	Sun et al. (2004a)
0.5 M NaOH and 0.5 M H_2O_2	0.01	0.01	0.35	0.02	0.011	0.01	Sun et al. (2004a)
Lignin fraction solubilized (80% dioxane solution with 0.05 M HCl–85 °C/4 h)	1.71	11.14	17.66	3.49	0.55[b]	0.55[b]	Sun et al. (2011)
Dewaxed bagasse with 1 M NaOH at 40 °C/18 h	0.007	0,018	0.070	0.018	0.70	0.42	Xu et al. (2005)

[a]*p*-Hydroxybenzoic acid
[b]As total cinnamic acid

6.3 Genomics Applied to the Study of Lignocellulosic Fungi

Whole-genome sequence analysis for lignocellulosic fungi is enabling identification of genes and isoforms encoding hydrolytic enzymes and facilitating downstream strain modification through genetic engineering. The first major milestone in fungal genomics was the sequencing of the yeast genome *Saccharomyces cerevisiae* (Goffeau et al. 1996). In the case of lignocellulosic fungi, the genomes of numerous species have now been completely sequenced or re-sequenced, contributing towards an understanding of lignocellulosic bioprocesses. Greater knowledge will enable improved hydrolytic enzyme cocktail development and downstream biotechnological applications.

Trichoderma is an important filamentous fungal genus for industrial bioprocesses, with numerous species capable of secreting cellulosic enzymes at high concentrations. *T. reesei* (teleomorph *Hypocrea jecorina*) has been the focus of whole-genome sequencing and re-sequencing projects for further understanding of lignocellulose degradation. The complete genome sequence of *T. reesei* was published by Martinez et al. (2008), with 34.1 Mb of sequence distributed over seven chromosomes and a total of 9143 gene models. Analysis of the *T. reesei* genome identified 200 genes that encode for glycosyl hydrolases, 103 glycosyl transferases, 36 carbohydrate-binding modules, and 3 polysaccharide lyases, including 7 genes encoding cellulases and 16 genes encoding hemicellulases (Martinez et al. 2008).

T. virens (teleomorph *Hypocrea virens*) and *T. atroviride* (teleomorph *Hypocrea atroviridis*) are also promising species regarding holocellulosic enzyme secretion. Whole-genome sequences have been completed for both species, with *T. virens* and *T. atroviride* genome sizes of 38.8 Mb and 36.1 Mb, respectively. As well as possessing greater genome sizes than *T. reesei*, the total numbers of predicted genes in these two species also outnumber those in *T. reesei*, with 12,428 genes predicted in *T. virens* and 11,865 in *T. atroviride*. A total of 260 genes in the *T. virens* genome and 257 in *T. atroviride* are predicted to be involved in the hydrolysis of cellulose and hemicelluloses, again outnumbering the total within these categories in *T. reesei* (Kubicek et al. 2011). Surprisingly, the *T. reesei* genome contained the lowest recorded number of holocellulase-encoding genes when compared with other fungal genomes, comprising a total of 130 genes located in 25 clusters for cellulose and hemicellulose degradation (Martinez et al. 2008). Additional *Trichoderma* species that are also promising for the production of hydrolytic enzymes include *T. longibrachiatum* and *T. harzianum*. Phylogenetically close to *T. reesei*, *T. longibrachiatum* has the smallest reported genome within the genus, totaling 32.24 Mb and 10,792 predicted genes. *T. harzianum*, by contrast, has the largest known genome size to date within the genus, totaling 40.98 Mb in size, with 14,095 predicted genes (Srivastava et al. 2014).

Fungal species of the genus *Aspergillus* are also promising secretors of enzymes that degrade lignocellulose, especially hemicellulose. *A. nidulans* (teleomorph *Emericella nidulans*), historically considered the model species for the genus, with the complete elucidation of its sexual cycle, has been the focus of whole-genomic sequencing (Galagan et al. 2005). The species *A. oryzae* and *A. niger* are classified as GRAS status (Generally Regarded as Safe), with wide employment in industrial applications (Machida et al. 2005; Pel et al. 2007). Whole-genome sequences for these two species are available in the genome databases at the Broad Institute (Broad-Institute 2015), the National Center for Biotechnology Information (NCBI 2015), and the *Aspergillus* Genome Database (AspGD 2015).

The *A. nidulans* genome, with a total size of 30.06 Mb and 9396 predicted genes, contrasts with the larger genomes of *A. oryzae* and *A. niger*, with respective genome sizes at 37.12 and 37.2 Mb, and gene models totaling 12,336 and 11,200 (Machida et al. 2005; Pel et al. 2007). From 186 predicted CAZy-encoding genes in *A. nidulans*, 159 genes have been reported to group in the GH family, 19 in the PL family, and 8 genes grouping in the CE family. In the case of *A. oryzae* and *A. niger*, 217 and 171 CAZy-encoding genes have been predicted, respectively, with 183 grouped in the GH family, 20 in the PL family, and 14 in the CE family for *A. oryzae* and 155, 8, and 8 genes grouped in families GH, PL, and CE, respectively, for *A. niger*. Comparative analyses of *A. oryzae*, *A. nidulans*, and *A. flavus* suggest that a number of mechanisms have been responsible for the differences in genome size and gene numbers among *Aspergillus* species, with genome and segmental duplication having resulted in genome size differences, and horizontal gene transfer shaping genomes between donor and recipient strains (Gibbons and Rokas 2013; Khaldi and Wolfe 2008; Mallet et al. 2010). Whole-genome sequences have also been generated and annotated for *A. terreus*, *A. flavus*, and *A. fumigatus*. *A. terreus* has a genome size of 29.33 Mb with 10,406 predicted genes (Broad-Institute 2015). By contrast, the *A. flavus* genome is estimated to be similar in size to *A. oryzae*, with 36.8 Mb coding 12,424 predicted genes (Gibbons and Rokas 2013). Similar to *A. nidulans*, *A. fumigatus* has a genome size of 29.2 Mb, encoding 9906 predicted genes (Nierman et al. 2005; Payne et al. 2006). A comparison between these two genomes shows that *A. fumigatus* also has a similar number of CAZy-encoding genes to *A. nidulans*, with 155 genes clustered in the GH family, 8 genes grouped in PL family, and 8 genes grouped in the EC family (Pel et al. 2007).

6.3.1 Functional Genomics Applied to the Study of Lignocellulosic Fungi

Functional genomics, or transcriptomics, which investigates the transcribed portion of the genome, has applications in both gene function identification and gene expression analysis. Large-scale comparative transcriptome analysis enables the identification of differentially expressed genes that are involved in conferring or regulating essential characteristics for the development and adaptation of a microorganism to its environment (Mardis 2008). Although filamentous fungi have been widely exploited for enzyme production under simultaneous saccharification and fermentation conditions (SSF), little research has been conducted on gene expression control mechanisms under SSF. While studies have shown that the expression of certain genes involved in the production of hydrolases is controlled at the transcriptional level, events following transcription and translation are less well documented in fungi (Lubertozzi and Keasling 2009). Applications of transcriptomics in fungi in relation to identification of genes involved in the production and regulation of hydrolytic enzymes have been the focus of considerable research for downstream application in the enzyme industries. In the late 1990s, the first transcriptome analyses in target fungi such as *Trichoderma* and *Aspergillus* focused on large-scale sequencing of expressed sequence tags (ESTs). In 2002, the transcriptome of *T. reesei* was analyzed in relation to glucose-dependent expression regulation, with 1151 unigenes identified among a set of 2835 ESTs (Chambergo et al. 2002). A similar study was conducted on *A. oryzae*, where from approximately 2000 selected ESTs, microarray analyses revealed that transcripts in cultures supplemented with glucose are highly dependent on the glycolytic pathway (Maeda et al. 2004). Since then, the total number of available ESTs deposited in the NCBI GenBank database has increased extensively. In the case of *Aspergillus* species which are the focus of hydrolytic enzyme production, 19,830 ESTs are currently available for *A. oryzae*, 16,394 for *A. nidulans*, 14,195 for *A. niger*, and 10,415 for *A. terreus*. For *Trichoderma* species, a more expressive number of ESTs are available in GenBank, reflecting the long-term recognized biotechnological importance of this genus. A total of 44,966 ESTs are available for *T. reesei*, 35,475 for *T. virens*, 35,125 for *T. atroviride*, 19,914 for *T. harzianum*, and 1799 ESTs for *T. longibrachiatum*.

Transcriptome analyses for lignocellulolytic species of *Trichoderma* and *Aspergillus* have focused on understanding the dynamics of genes involved in the synthesis of enzymes of industrial value. Transcription profiles after growth on specific culture media have contributed to the identification of novel differentially expressed genes. For example, gene expression analysis in *T. reesei* strains during limiting culture conditions with cellulose, lactose, and sophorose enabled characterization of 5131 predicted unigenes, with novel genes identified that encode previously unknown endoglucanases (cel74a, cel61b, and cel5b), β-glucosidases (cel1b, cel3b, cel3c, cel3d, and cel3e), hemicellulases (axe2 and ABF2), and genes containing cellulose-binding domains (CIP1 and CIP2). Northern blot and microarray analyses indicated that these novel genes are highly regulated in the presence of cellulose and sophorose (Foreman et al. 2003).

Similarly, transcriptomic analysis in *A. niger* when cultivated on media supplemented with glucose, maltose, xylan, xylose, sorbitol, and lactose enabled the identification of 5108 unigenes from 12,820 ESTs. A subsequent Gene Ontology (GO) classification of those genes revealed 2549 predicted protein products of which 32% showed a hydrolytic activity (Semova et al. 2006).

Next-generation sequencing (NGS) technologies applied to transcriptome analysis enable large-scale and precise characterization of unigenes and their expression (Mardis 2008). Illumina RNA-Seq is a shotgun method appropriate for transcriptome sequencing, where cDNA is fragmented mechanically to produce small overlapping fragments that can cover the entire transcriptome. This approach allows identification of novel genes and isoforms involved in protein production under contrasting conditions. In addition, this massively parallel sequencing technology offers a sensitivity appropriate for characterization of all transcripts, including even those with low expression levels (Garber et al. 2011). In this context, NGS technologies may be applied to identifying genes involved in cell wall degradation pathways as well as understanding transcriptional control mechanisms in lignocellulolytic fungi with biotechnological potential.

Genes required for growth and enzymatic activity on polymers such as cellulose have been analyzed by NGS in the model organism *Neurospora crassa*. Two uncharacterized zinc binuclear cluster transcription factors, crl1 and crl2, were identified as responsible for the induction of the major genes encoding cellulases and hemicellulases (Coradetti et al. 2012; Znameroski et al. 2012). Similarly, in the anaerobic rumen fungus *Neocallimastix patriciarum*, NGS approaches allowed the identification of genes involved in the degradation of different lignocellulosic materials (Wang et al. 2011). In the basidiomycete fungus *Phanerochaete carnosus*, NGS-based analyses during culture on different wood substrates revealed that the abundance of transcripts encoding enzymes with activity in lignin degradation was greater than those coding for enzymes involved in degradation of other carbohydrates (MacDonald et al. 2011).

NGS-based transcriptional profiling of genes encoding enzymes with carbohydrate activity in *Myceliophthora thermophila* and *Thielavia terrestris* was also investigated following growth on barley straw, alfalfa straw, and glucose. Upregulation of genes involved in lignocellulose degradation was observed following exposure to polysaccharides derived from these biomass substrates (Berka et al. 2011). Delmas et al. (2012) reported the application of SOLiD® RNA-Seq NGS to analyze the transcriptome of *A. niger*. The study found that approximately 19% of the total mRNA is involved in gene expression related to enzymes that degrade plant cell wall after 24-h exposure to wheat straw. This is a considerable enrichment of CAZy group genes, given that they represent only 2.5% of the coding genome. The numerous categories of expressed CAZy genes during exposure to wheat straw reflected the complexity of the carbohydrates present in the substrate. Interestingly, approximately 65% of the mRNA from CAZy groups originated from genes that encode only five enzyme families, namely GH7, 11, 61, and 62 (cellobiohydrolases, xylanases, α-arabinofuranosidases, and polysaccharide monooxygenases, respectively) and CE1 (acetyl xylan esterases). Based on transcript abundance, these five categories of encoded enzymes were considered to provide the complete spectrum of activities for straw degradation. Catabolic repression by glucose addition to wheat straw cultures was also reported to repress CAZy gene expression after just 5 h, with CAZy gene mRNA reduced to just 1% of the total RNA. Analysis has shown that the predominant induction model in fungal systems proposes a basal expression of small amounts of hydrolases that initiate degradation of complex polysaccharides, thus producing inducing compounds which provoke a complete transcription response (Foreman et al. 2003).

Comparative studies of the transcriptome of different lignocellulolytic fungi have shown that as well as a repression of genes encoding hydrolytic enzymes in the presence of glucose, an induction of these genes occurs in the presence of compounds derived from the material present in the plant cell wall or its metabolic products. Analysis of Illumina RNAseq-derived transcriptome sequences of *T. reesei* and *A. niger*, grown in the presence of glucose or wheat straw as a carbon source, revealed that *T. reesei* uses an enzyme machinery similar to that of *A. niger*, suggesting that both microorganisms employ conserved strategies for the depolymerization of lignocellulosic substrates. In the case of *T. reesei*, after 48-h growth on glucose, mRNA of genes encoding CAZy enzymes represented only 1.14% of the total RNA. By contrast, following 24-h incubation in the presence of wheat straw, genes that expressed mRNA for GH families (β-glucosidase, xylanase, and cellobiohydrolase CEL7A), AA9 families (auxiliary oxidoreductases), and EC families (acetyl xylan esterases) represented around 65% of the expressed CAZy genes. Similar changes in expression of CAZy enzyme-encoding genes in relation to carbon source were observed in *A. niger* (Delmas et al. 2012; Ries et al. 2013). Interestingly, recent comparative secretome analysis on lignocellulosic biomass has indicated that depolymerization can occur via different mechanisms in each of these fungal species (Borin et al. 2015).

Recently, the CAZy repertoire in *T. reesei* based on computational and manual approaches was revised and expanded, which resulted in a total of 201 genes encoding glycosyl hydrolases, 22 genes encoding carbohydrate esterases, and 5 genes encoding polysaccharide lyases. Most of the genes that express cellulases and hemicellulases characterized in *T. reesei* are regulated by the type of carbon source available, ensuring the production of only hydrolytic enzymes necessary for degradation of that particular substrate. Analysis of transcription factors in *T. reesei* that are involved in expression regulation of genes encoding cellulases and hemicellulases has also been reported. These include positive regulators such as XYR1, ACE2, and the complex HAP2/3/5, as well as negative regulators such as ACE1 and the catabolic repressor CRE1 (Häkkinen et al. 2012). The transcription factor XYR1 (xylanase regulator 1) is recognized as a main transcriptional activator of genes encoding hydrolases such as xyn1, xyn2, bxl1, cbh1, cbh2, and egl1 (Portnoy et al. 2011). In addition, the analysis showed that XYR1- and CRE1-derived regulation of expression of genes encoding cellulases and xylanases is carbon source dependent (Castro et al. 2014).

Castro et al. (2014) recently investigated the *T. reesei* transcriptome following the growth on cellulose, sophorose, and glucose media cultures. Analysis of NGS data enabled hierarchical clustering of 2060 differentially expressed genes when comparing expression on cellulose against glucose, sophorose against cellulose, and glucose against sophorose. Three possible regulators were identified, with 123 genes modulated by cellulose, 154 genes modulated by sophorose, and 402 genes modulated by glucose. Comparison of the top ten most differentially expressed genes following growth on the three carbon sources was conducted to evaluate the specific regulation of each carbon source. The ten most upregulated genes in cellulose included glycosyl hydrolases family members (GH5, GH31, GH16), carbohydrate esterase (EC 5), a major facilitator superfamily permease (MFS), and five proteins with unknown function. For sophorose, one hydrolase (GH76), four oxidoreductases, two MFS, and three proteins with unknown function were identified. As expected for growth on glucose, the top ten most differentially expressed genes were not related to hydrolytic enzyme-encoding genes.

6.3.2 Functional Genomics Analysis of Lignocellulosic Fungi on Sugarcane Bagasse

Considering the high cost of the major industrial enzymes employed today in the hydrolysis of sugarcane bagasse (cellulases and hemicellulases), together with limitations in their efficiency, an increased understanding of novel enzymes and mechanisms controlling their expression in candidate lignocellulosic fungi is important for downstream efficient biotechnological enzyme production (de Souza et al. 2014).

A number of studies have been conducted to identify genes and transcriptional control mechanisms in lignocellulolytic fungi hydrolyzing sugarcane bagasse. A microarray-based analysis of 3700 genes from *A. niger* grown on this biomass as a carbon source enabled the identification of 18 genes encoding cellulases which were upregulated in response to sugarcane bagasse compared to fructose. A similar response was observed with 21 genes encoding hemicellulases. Seven genes encoding predicted transporters also showed an increased expression in response to sugarcane bagasse and a repression in the presence of glucose, suggesting that these genes are xylose transporters (Souza et al. 2011). Continuing this work, this group examined differential expression of genes in *A. niger* that are regulated by the transcriptional activator XlnR and its close homolog, AraR, which both control the main cellulolytic and hemicellulolytic system in response to growth on sugarcane bagasse. Global gene expression analysis revealed a total of 1024, 439, 109, and 176 genes that were significantly and exclusively upregulated in *A. niger* wild-type, ΔxlnR, ΔaraR, and ΔaraRxlnR strains, respectively, while 1553, 770, 555, and 246 genes were exclusively downregulated. The considerable numbers of differentially expressed genes observed indicated the involvement of transcriptional regulators controlling metabolism in response to this complex plant substrate (Souza et al. 2013).

To identify genes involved in degradation of cellulose and hemicellulose together with accessory genes involved in the depolymerization of the biomass, Horta et al. (2014) characterized the transcriptome of the fungus *T. harzianum* grown on sugarcane bagasse using NGS RNA-seq technology. A total of 487 CAZymes were identified from the transcriptome, with 23 genes encoding glycosyl hydrolase family proteins involved in the depolymerization of cellulose and a further 22 glycoside hydrolase-encoding genes involved in the hemicellulose breakdown. A total of 164 contigs were also classified as potentially related to the degradation of sugarcane biomass. A systematic synergism between different enzyme classes for specific metabolic pathways in *T. harzianum* was also identified. These pathways included metabolism of various sugars associated with the depolymerization of this biomass.

Similarly, functional genomics analyses of a number of promising lignocellulolytic fungi are being conducted at the University of Brasilia, following growth on abundant organic residues that include sugarcane bagasse, soybean hulls, and cotton louse. RNA-seq analysis of a strain of *A. tamarii* cultivated on sugarcane bagasse as a carbon source enabled the identification of a total of 207 CAZy-encoding genes in liquid culture state, and 222 genes in semisolid culture, from a total of 7126 genes expressed by the fungus across different cultivation treatments (Midorikawa, G.E.O., unpublished). Similarly, RNA-seq analysis of a local strain of *A. terreus* grown on sugarcane bagasse showed that 102 CAZy-encoding genes were expressed during liquid culture state, from a total of 6200 expressed genes (Correa, C. L., unpublished).

6.4 Enzymatic Attack on the Sugarcane Bagasse Structure

Enzymatic hydrolysis of plant cell wall is known to require an arsenal of enzymes, including cellulases, hemicellulases, and pectinases. A kinetic model for the interaction between cell wall components and a consortium of enzyme systems needs the analysis of several factors, including the involvement of different types of chemical linkages and the environment that surrounds the cell wall structure. Borin et al. (2015) showed that in both secretomes of *A. niger* and *T. reesei*, the most important enzymes related to culm in natura and pretreated bagasse degradation were cellobiohydrolases, endoglucanases, β-glucosidases, β-xylosidases, endoxylanases, xyloglucanases, and α-arabinofuranosidases. Moreover, both fungi produce more enzymes when grown in culm as a single carbon source. In addition, *A. niger* produced more enzymes (quantitatively and qualitatively) than *T. reesei*.

As a recalcitrant structure, the plant cell wall demands a broad range of exo- and endo-hydrolytic enzymes acting together for its complete degradation (Coughlan et al. 1993; de Souza et al. 2013). The complexity and heterogeneity of the holocellulose structures demand the synergistic action of main- and side-chain attacking enzymes with different specificities for its complete hydrolysis (Andreaus et al. 2008; Moreira et al. 2012; Buckeridge and de Souza 2014). Synergism is only detected when the total products formed by two or more enzymes acting together exceed the arithmetic sum of the products formed by the action of each enzyme individually. Enzyme systems with low substrate specificity might act in synergism with enzymes with strict substrate specificity, leading to a more efficient catalytic process (Buckeridge 2010). Within this context, enzymes that cut specific sites in the plant cell wall are also important tools for understanding the structure and function of the cell wall (Siqueira and Filho 2010; Buckeridge and de Souza 2014).

Specifically for sugarcane bagasse, two side-chain enzymes appear as key enzymes to its full degradation (Fig. 6.1): feruloyl esterase, which is responsible for cleaving the ester link between the polysaccharides in the main chain of xylans or pectins, and α-arabinofuranosidase that catalyzes the hydrolysis of terminal α-L-1,2-, α-L-1,3-, and α-L-1,5-arabinofuranosyl residues of poly- and oligosaccharides (Moreira et al. 2011).

6.4.1 Feruloyl Esterases

Feruloyl esterases are also known as ferulic acid esterases, cinnamoyl esterases, and cinnamic acid hydrolase (FAEs, E.C. 3.1.1.73), a subclass of the carboxylic acid esterases (E.C. 3.1.1.1). These enzymes have gained importance in the biofuel, medicine, and food industries due to their capability of hydrolyzing carbohydrate esters in wood polymers, and synthesizing high added value molecules through esterification and transesterification reactions (Udatha et al. 2012; Crepin et al. 2003).

An enzyme that hydrolyzes ester linkages of ferulic acid from starch-free wheat bran and from a soluble ferulic acid-sugar ester, isolated from wheat bran, was first reported as a component of the cellulolytic and xylanolytic systems of *Schizophyllum commune* (Mackenzie and Bilous 1988). Since the 1990s, a broad range of FAEs have been purified and characterized from a wide range of microorganisms, including bacteria and fungi. Many enzymes from the genera *Aspergillus*, *Clostridium*, and *Penicillium* have been studied (de Vries et al. 2002; Topakas et al. 2007; Moreira et al. 2011). FAEs from different microbial sources present great variations in physiochemical properties, concerning to their molecular masses, pI, optimum pH and temperatures, and kinetic parameters (revised by Wong 2006).

FAEs are thought to be responsible for cleaving the ferulic acid bridges and eliminating the cross-links between polysaccharides and the lignin structure. These polysaccharides, which complex together with macromolecular complexes like lignin and cellulose, modulate the plant cell wall material. To date, feruloyl esterases are also thought to be key in the enzymatic synthesis of phenolic sugar esters (Moreira et al. 2011; Vafiadi et al. 2006; Wong 2006). FAEs act in synergy with main-chain-degrading enzymes, such as β-(1,4)-endoxylanases, to increase the release of ferulic acid from plant cell wall (Levasseur et al. 2006). FAEs also enhance the release of glucose, so a higher release of glucose was obtained when cellulase, xylanase, and feruloyl esterase were used in the enzyme mixture to hydrolyze wheat straw for bioethanol production (Tabka et al. 2006).

Feruloyl esterases occur either as a non-modular enzyme (Crepin et al. 2003), or they can be part of multimodular protein structures, as cellulosomes (Prates et al. 2001; Blum et al. 2000). Moreover, some FAEs contain cellulose-binding modules (CBMs) (Kroon et al. 2000). The increase in the catalytic efficiency of FAEs has been reached by the fusion of CBM in a chimeric enzyme composed of FAE and endo-xylanase (Levasseur et al. 2005). As mentioned above, the carbon source has an important effect on the enzymes produced during microorganism cultivation. Wheat bran seems to be the best inducer for FAE production, and the washing of the wheat bran favors further enzymatic production (Braga et al. 2014).

Feruloyl esterases have been shown to be important enzymes in sugarcane bagasse degradation. The supplementation of a commercial enzymatic preparation with a crude extract from *A. oryzae* containing ferulic acid esterase increased by 36% the conversion of cellulose in pretreated sugarcane bagasse (Braga et al. 2014). Moreover, a synergistic effect was observed in saccharification of steam-pretreated sugarcane bagasse by a mixture of crude extracts from *T. reesei* and *A. awamori*, the latter being a promising producer of ferulic acid esterase (Gottschalk et al. 2010). A mixture of cellulases and xylanases was supplemented with feruloyl esterases AnFaeA from *A. niger*, TsFaeC from *Talaromyces stipitatus*, and NcFaeD from *N. crassa*, to compare their potential to enhance total solubilization of sugarcane bagasse. Only the type C feruloyl esterase (TsFaeC) was able to release *p*-coumaric acid (pCA) from the natural substrate. However, no increment in the release of glucose was observed, which could also indicate that, despite phenolic acid substitution, TrCel7B and TmXyn10A were capable of fragmenting xylans that may limit access to cellulose microfibrils. A moderate improvement in xylose release could be observed, especially by TsFaeC, which indicates that solubilization of hydroxycinnamoyl groups from xylan or xylooligosaccharides enhanced saccharification of xylan to xylose monomers. The effect of feruloyl esterases was less clear. The three feruloyl esterases, types A, C, and D (see below the definitions for each type), had distinct product profiles on non-pretreated sugarcane bagasse substrate, indicating that bagasse could function as a possible natural substrate for FAE activity measurements (Varnai et al. 2014).

6.4.1.1 Classification of FAEs

Feruloyl esterases are currently classified as carbohydrate esterases (CE) of the family 1, based on their amino acid sequence (Henrissat 1991; CAZy 2015). In order to provide a more elaborated classification, FAEs have been grouped according to alignments of sequences or domains. The outcome of this genetic comparison has supported substrate specificity data and allowed feruloyl esterases to be subclassified into four types (A–D) based on similarities in esterase activity profiles against synthetic methyl esters. The intra- and intergroup protein sequence identities of the enzymes indicate an evolutionary relationship between feruloyl esterases, acetyl xylan esterases, and certain lipases (Crepin et al. 2004; Wong 2006). Each feruloyl esterase has its own specificity concerning the release of specific cinnamic acids (Benoit et al. 2006).

Type A feruloyl esterases tend to be induced during growth on cereal-derived substrates and show a preference for the phenolic moiety of the substrate that contains methoxy substitutions, especially at carbon(s) 3 and/or 5, as occurs in ferulic and sinapic acids (Crepin et al. 2004). These FAEs have sequences related to those of lipases and can hydrolyze synthetic ferulate dehydrodimers. Examples of this group of enzymes include *A. niger* FAE-A (AnFaeA) (Wong 2006). Type B feruloyl esterases are preferentially secreted during growth on sugar beet pulp and release ferulic acid ester linked to either C-2 of feruloylated arabinose or C-6 of feruloylated galactose residues (Kroon and Williamson 1996). However, they are unable to release the dimeric forms of ferulate. The type B feruloyl esterases show a preference for the phenolic moiety of the substrate that contains one or two hydroxyl substitutions, as found in *p*-coumaric and caffeic acids, respectively (Crepin et al. 2004). These enzymes show sequence similarities to carboxylic esterase family

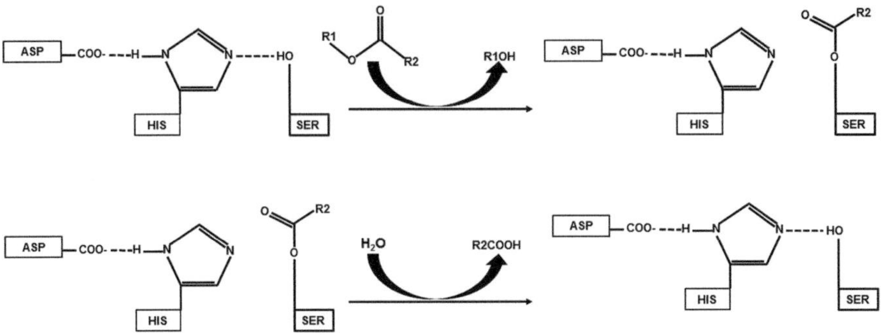

Fig. 6.2 The general mechanism of feruloyl esterase action. The reaction is processed in two steps: in the first one, an acyl intermediate is linked to the serine residue, while the portion of the molecule containing the radical R1 is released. In the second step, a hydrolysis reaction is responsible for the second moiety release. Considering that the molecule in the figure is a feruloylated arabinosyl residue, the first and second products would be an arabinose molecule (harboring radical R1) and ferulic acid, respectively

1—acetyl xylan esterase. *P. funiculosum* FAE-B (Wong 2006) and *N. crassa* FAE-I belong to this group (Crepin et al. 2003). Regarding specificity against synthetic substrates, both types of feruloyl esterases are active against methyl ferulate and methyl *p*-coumarate. Only type A is active against methyl sinapate and only type B is active against methyl caffeate. Type A feruloyl esterase appears to prefer hydrophobic substrates with bulky substituents on the benzene ring (Crepin et al. 2004; Wong 2006). A third type and fourth type of feruloyl esterase with broad specificity against synthetic hydroxycinnamic acids were also reported, denominated types C and D. They both hydrolyze the four methyl esters of hydroxycinnamic acids which are generally used as model substrates (methyl ferulate, methyl sinapinate, and methyl *p*-coumarate and methyl caffeate) but not diferulic compounds (Benoit et al. 2006; Wong 2006). Types C and D show sequence similarities to chlorogenate esterase and xylanase, respectively. However, only type D feruloyl esterases are able to hydrolyze dimers. Type C includes *A. niger* FAE-B (AnFaeB) and *Talaromyces stipitatus* FAE-C, while type D enzymes include *Piromyces equi* EstA and *Celluvibrio japonicus* esterase D (Wong 2006).

6.4.1.2 Catalytic Mechanism of FAEs

FAEs possess a common characteristic feature of the Ser-His-Asp catalytic triad mechanism (Fig. 6.2). Some variations in amino acid sequences forming surface loops and additional domains allow them to accommodate different aromatic substrates. In this case, serine acts as a nucleophile, histidine as the general acid–base, and the aspartic acid helps to orient the histidine residue and further neutralize the charge that forms on histidine during the catalytic process. Therefore, the protonation states or the net charge of aspartic acid and histidine residues are very important for maintaining the activity of FAEs (Udatha et al. 2012; Ekici et al. 2008).

The crystal structures of FAEs from *A. niger* show that these enzymes have a common α/β hydrolase fold and a catalytic triad Ser-His-Asp also common in lipases (Hermoso et al. 2004; McAuley et al. 2004; Faulds et al. 2005). A ferulate esterase domain, which is a component of a multimodular xylanase, called Xyn10B from *C. thermocellum*, also displays the α/β hydrolase fold and a catalytic triad Ser-His-Asp (Prates et al. 2001). The same catalytic triad was observed for an FAE from *A. oryzae* (Zeng et al. 2014).

6.4.2 Arabinofuranosidases

As mentioned before, arabinoxylan is the most abundant hemicellulose in annual plants such as sugarcane bagasse, with a backbone of β-(1,4)linked xylosyl residues substituted with arabinose (Xu et al. 2005). The α-L-arabinofuranosyl (α-L-Araf) residues are part of many plant cell wall polysaccharides, including arabinoxylan, and the pectic polysaccharides arabinogalactan and arabinan (Kim 2008). α-L-Arabinofuranosidases (α-L-AFases, EC 3.2.1.55) are accessory enzymes that cleave α-L-arabinofuranosidic linkages at the nonreducing ends of arabinose-containing polysaccharides and play a key role in the biodegradation of hemicellulose (Saha 2000). They catalyze the hydrolysis of terminal α-L-1,2-, α-L-1,3-, and α-L-1,5-arabinofuranosyl residues, but also residues from hemicellulosic and pectic polysaccharides (branched arabinans, debranched arabinans) and heteropolysaccharides (arabinogalactans, arabinoxylans, arabinoxyloglucans, glucuronoarabiloxylans, etc.). They act synergistically with other hemicellulases and pectic enzymes for the complete hydrolysis of hemicelluloses and pectins (Rye and Withers 2000; Shallom et al. 2002; Saha 2000; Shofiqur Rahman et al. 2003; Moreira et al. 2011). α-L-AFases perform the hydrolysis of α-glycosidic bonds in a highly accelerated rate (by more than 107-fold), making them one of the most efficient catalysts known (Shallom et al. 2002).

A broad range of α-L-AFases have been purified, and great diversity in the physicochemical properties, such as pH, temperature, thermostability, and molecular mass, was noted among the enzymes from plant, fungal, and bacterial sources. Microbial α-L-AFases exhibit a diversity of pH and temperature activities, and present monomeric or dimeric form (Revised by Saha 2000). Arabinose, a competitive inhibitor (Ki 16.4 mM) of Ara II, purified from *P. capsulatum*, had no effect on Ara I (purified from the same fungus) activity at concentrations of up to 40 mM (Filho et al. 1996). Arabinose also acted as a noncompetitive inhibitor with Ki of 38.4 mM over an arabinofuranosidase (Abfa) from *Aureobasidium pullulans* (de Wet et al. 2008). Moreover, α-L-AFases possessing β-xylosidase activity or xylanases with α-L-arabinofuranosidase activity have also been described (Mai et al. 2000; Tateishi et al. 2014; de Wet et al. 2008). The relaxed specificity observed in these bifunctional enzymes is attributed to the structural features of catalytic sites, which allows the accommodation of different substrates (Lee et al. 2003). α-L-AFases have been classified into seven different GH families, namely GH2, GH3, GH43, GH51, GH54, GH62, and GH93 (CAZy 2015; Henrissat and Bairoch 1993).

As well as feruloyl esterases, α-arabinofuranosidases are essential enzymes for the complete deconstruction of sugarcane bagasse. The addition of α-L-Araf and pectinase to a cellulase cocktail from *T. harzianum* was responsible for an increase of 116% in the total reducing sugar released during pretreated sugarcane bagasse hydrolysis (Delabona et al. 2013). A synergistic effect of a xylanase (XynC) and an α-arabinofuranosidase (AbfB) was observed in the production of xylooligosaccharides after hydrolysis of pretreated sugarcane bagasse. These results suggest that the synergistic improvement of xylooligosaccharide production is correlated by the action of AbfB on xylooligosaccharides having an arabinofuranose residue (Gonçalves et al. 2012). The efficiency of recombinant α-arabinofuranosidases, endo-xylanases, and β-xylosidases for sugarcane hydrolysis in several combinations of these enzymes were also evaluated, and it was found that synergic action of α-arabinofuranosidases and endo-xylanases led to the production of higher amounts of xylose liberated from xylan backbone. This indicates the importance of the action of removal of lateral arabinofuranosyl residues to allow access of endo-xylanase to xylose main chain (Goldbeck et al. 2014).

6.4.2.1 Catalytic Mechanism of AFses

Arabinofuranosidases belonging to GH43 and GH62 families possess conserved catalytic residues of Asp-Asp-Glu. An α-arabinofuranosidase, ScAraf62A from *Streptomyces coelicolor* has the catalytic triad Asp202-Asp309-Glu361, and no enzyme activities were observed for the mutants Asp202 and Glu361 (Maehara et al. 2014). It was predicted that Asp and Glu behave as a general base and acid catalytic residues, respectively (Shallom et al. 2005).

It was predicted that an α-L-AFase belonging to the GH 51 family, isolated from *Geobacillus stearothermophilus* T-6 hydrolyzes glycosidic bonds using a two-step double-displacement mechanism or the retaining mechanism. In the glycosylation step of the reaction, the acid–base residues act as a general acid by protonation of the glycosidic oxygen and stabilizing the leaving group. In turn, the nucleophilic residue attacks the anomeric center of the scissile bond, forming a covalent glycosyl-enzyme intermediate. During the deglycosylation, the acid–base residue acts as a general base and activates a water molecule that attacks the anomeric center of the glycosyl–enzyme intermediate, liberating the free sugar with an overall retention of the anomeric configuration (Fig. 6.3). Both steps involve the formation of an oxocarbenium ion-like transition state (Shallom et al. 2002).

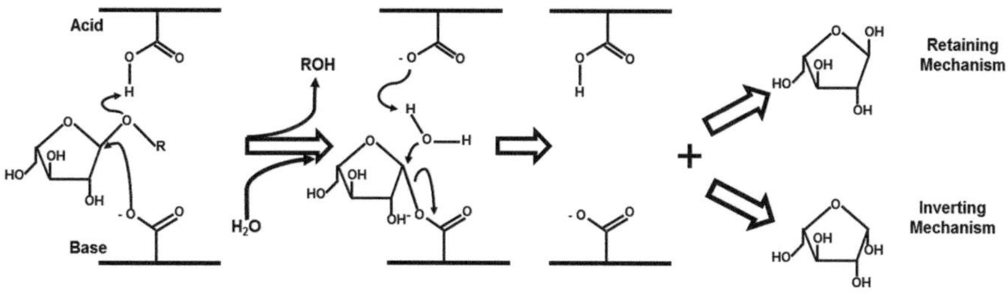

Fig. 6.3 The reaction mechanism of α-L-arabinofuranosidases. Glutamic and aspartic acids act as acid and base in the hydrolysis mechanism. α-L-Arabinofuranosidases may release an arabinofuranosyl residue with both inverted and conserved positions of the anomeric carbon. *R* represents the group to which the arabinofuranosyl residue is associated

6.5 Conclusion and Perspectives

Bagasse is an important residue generated during sugarcane processing, and its accumulation is a source of many environmental problems. However, its utilization as a feedstock for second-generation bioethanol production depends on efficient enzyme cocktails containing accessory enzymes in addition to main-chain cleaving enzymes.

For sugarcane bagasse, the synergic contribution of ferulic acid esterases and arabinofuranosidase seems to have great importance for the efficient degradation of the biomass. If added to enzymatic cocktails, they may allow the use of reduced protein loads, so reaching a more efficient biomass hydrolysis along with low costs.

Filamentous fungi are promising organisms for the production of broad-spectrum hydrolytic enzymes. The characterization of the transcriptome of diverse lignocellulolytic fungi is an important approach for the identification of novel genes involved in the regulation and expression of these enzymes.

As gene expression changes as a result of DNA methylation, chromatin remodeling and RNA interference have also recently been observed in filamentous fungi (Aghcheh and Kubicek 2015); potential also exists in certain epigenetic mechanisms for fungal strain improvement for biotechnological applications. Similarly, events following transcription and translation have yet to be well characterized in lignocellulolytic fungi, with current proteomic approaches applicable for high-resolution elucidation of biomass degradation mechanisms (Borin et al. 2015).

Increased knowledge of genes, regulation processes, and overall hydrolytic mechanisms present in fungi that are involved in the degradation of organic residues such as sugarcane bagasse will contribute to the second-generation bioethanol industry through the identification and development of improved strains capable of hydrolyzing plant biomass to simple fermentable sugars. Increased enzymatic efficiencies and improved hydrolytic enzyme cocktail development will contribute to reducing cellulosic bioethanol production costs, given that a main bottleneck in the production process is the recalcitrance of the plant cell wall.

Acknowledgements The authors acknowledge the financial support from the Brazilian Research Council (CNPq), Coordination for the Improvement of Higher Level Personnel (CAPES), the Foundation for Research Support of Federal District (FAPDF), and the National Institute of Science and Technology of Bioethanol.

References

Aghcheh RK, Kubicek CP (2015) Epigenetics as an emerging tool for improvement of fungal strains used in biotechnology. Appl Microbiol Biotechnol 99:6167–6181

Andreaus J, Filho EXF, Bon EPDS (2008) Biotechnology of holocellulose-degrading enzymes. In: Hou CT, Shaw J (eds) Biocatalysis and bioenergy. Wiley, Hoboken

ASPGD (2015) Aspergillus genoma database [online]. Accessed 23 Mar 2015

Battaglia E, Benoit I, van den Brink J, Wiebenga A, Coutinho PM, Henrissat B, de Vries RP (2011) Carbohydrate-active enzymes from the zygomycete fungus Rhizopus oryzae: a highly specialized approach to carbohydrate degradation depicted at genome level. BMC Genomics 12:38

Benoit I, Navarro D, Marnet N, Rakotomanomana N, Lesage-Meessen L, Sigoillot J-C, Asther M, Asther M (2006) Feruloyl esterases as a tool for the release of phenolic compounds from agro-industrial by-products. Carbohydr Res 341:1820–1827

Berka RM, Grigoriev IV, Otillar R, Salamov A, Grimwood J, Reid I, Ishmael N, John T, Darmond C, Moisan M-C (2011) Comparative genomic analysis of the thermophilic biomass-degrading fungi Myceliophthora thermophila and Thielavia terrestris. Nat Biotechnol 29:922–927

Beukes N, Chan H, Doi RH, Pletschke BI (2008) Synergistic associations between Clostridium cellulovorans enzymes XynA, ManA and EngE against sugarcane bagasse. Enzyme Microb Technol 42:492–498

Blum DL, Kataeva IA, Li X-L, Ljungdahl LG (2000) Feruloyl esterase activity of the Clostridium thermocellum cellulosome can be attributed to previously unknown domains of XynY and XynZ. J Bacteriol 182:1346–1351

Borin GP, Sanchez CC, de Souza AP, de Santana ES, de Souza AT, Leme AFP, Squina FM, Buckeridge M, Goldman GH, Oliveira JV (2015) Comparative secretome analysis of Trichoderma reesei and Aspergillus niger during growth on sugarcane biomass. PLoS One 10(6):e0129275

Braga CMP, Da Silva Delabona P, da Silva Lima DJ, Paixão DAA, da Cruz Pradella JG, Farinas CS (2014) Addition of feruloyl esterase and xylanase produced on-site improves sugarcane bagasse hydrolysis. Bioresour Technol 170:316–324

Brienzo M, Ferreira S, Vicentim MP, de Souza W, Sant'anna C (2014) Comparison study on the biomass recalcitrance of different tissue fractions of sugarcane culm. Bioenergy Res 7:1454–1465

Broad-Institute (2015) Broad Institute [online]. Accessed 23 Mar 2015

Buckeridge MS (2010) Seed cell wall storage polysaccharides: models to understand cell wall biosynthesis and degradation. Plant Physiol 154:1017–1023

Buckeridge MS, de Souza AP (2014) Breaking the "Glycomic Code" of cell wall polysaccharides may improve second-generation bioenergy production from biomass. Bioenergy Res 7: 1065–1073

Castro LDS, Antoniêto ACC, Pedersoli WR, Silva-Rocha R, Persinoti GF, Silva RN (2014) Expression pattern of cellulolytic and xylanolytic genes regulated by transcriptional factors XYR1 and CRE1 are affected by carbon source in Trichoderma reesei. Gene Expr Patterns 14:88–95

CAZY (2015) Carbohydrate-active enzyme database [online]. http://www.cazy.org/. Accessed 15 Jan 2015

Chambergo FS, Bonaccorsi ED, Ferreira AJS, Ramos ASP, Ferreira JR, Abrahão-Neto J, Farah JPS, El-Dorry H (2002) Elucidation of the metabolic fate of glucose in the filamentous fungus Trichoderma reesei using expressed sequence tag (EST) analysis and cDNA microarrays. J Biol Chem 277:13983–13988

Chandel A, Antunes F, Silva M, da Silva S (2013) Unraveling the structure of sugarcane bagasse after soaking in concentrated aqueous ammonia (SCAA) and ethanol production by Scheffersomyces (Pichia) stipitis. Biotechnol Biofuels 6:102

Chandel AK, da Silva SS, Carvalho W, Singh OV (2012) Sugarcane bagasse and leaves: foreseeable biomass of biofuel and bioproducts. J Chem Technol Biotechnol 87:11–20

Coradetti ST, Craig JP, Xiong Y, Shock T, Tian C, Glass NL (2012) Conserved and essential transcription factors for cellulase gene expression in ascomycete fungi. Proc Natl Acad Sci 109:7397–7402

Coughlan MP, Tuohy MG, Filho EXF, Puls J, Claeyssens M, Vrsanska M, Hughes MM (1993) Enzymological aspects of microbial hemicellulases with emphasis on fungal systems. In: Coughlan MP, Hazlewood GP (eds) Hemicellulose and hemicellulases, Portland Press Research Monograph IV. Portland Press, Cambridge

Couturier M, Navarro D, Olivé C, Chevret D, Haon M, Favel A, Lesage-Meessen L, Henrissat B, Coutinho PM, Berrin J-G (2012) Post-genomic analyses of fungal lignocellulosic biomass degradation reveal the unexpected potential of the plant pathogen Ustilago maydis. BMC Genomics 13:57

Crepin VF, Faulds CB, Connerton IF (2003) A non-modular type B feruloyl esterase from Neurospora crassa exhibits concentration-dependent substrate inhibition. Biochem J 370: 417–427

Crepin VF, Faulds CB, Connerton IF (2004) Functional classification of the microbial feruloyl esterases. Appl Microbiol Biotechnol 63:647–652

de Souza AP, Grandis A, Leite DCC, Buckeridge MS (2014) Sugarcane as a bioenergy source: history, performance, and perspectives for second-generation bioethanol. Bioenergy Res 7:24–35

de Souza AP, Leite DC, Pattathil S, Hahn MG, Buckeridge MS (2013) Composition and structure of sugarcane cell wall polysaccharides: implications for second-generation bioethanol production. Bioenergy Res 6:564–579

de Vries RP, Vankuyk PA, Kester HCM, Visser J (2002) The Aspergillus niger faeB gene encodes a second feruloyl esterase involved in pectin and xylan degradation and is specifically induced in the presence of aromatic compounds. Biochem J 363:377–386

de Wet BJ, Matthew MK, Storbeck K-H, van Zyl WH, Prior BA (2008) Characterization of a family 54 α-L-arabinofuranosidase from Aureobasidium pullulans. Appl Microbiol Biotechnol 77:975–983

Delabona PDS, Cota J, Hoffmam ZB, Paixão DAA, Farinas CS, Cairo JPLF, Lima DJ, Squina FM, Ruller R, Pradella JGDC (2013) Understanding the cellulolytic system of Trichoderma harzianum P49P11 and enhancing saccharification of pretreated sugarcane bagasse by supplementation with pectinase and α-l-arabinofuranosidase. Bioresour Technol 131:500–507

Delmas S, Pullan ST, Gaddipati S, Kokolski M, Malla S, Blythe MJ, Ibbett R, Campbell M, Liddell S, Aboobaker A (2012) Uncovering the genome-wide transcriptional responses of the filamentous fungus Aspergillus niger to lignocellulose using RNA sequencing. PLoS Genet 8:e1002875

Ekici ÖD, Paetzel M, Dalbey RE (2008) Unconventional serine proteases: variations on the catalytic Ser/His/Asp triad configuration. Protein Sci 17:2023–2037

Ensinas AV, Nebra SA, Lozano MA, Serra LM (2007) Analysis of process steam demand reduction and electricity generation in sugar and ethanol production from sugarcane. Energ Conver Manage 48:2978–2987

Faulds CB, Molina R, Gonzalez R, Husband F, Juge N, Sanz-Aparicio J, Hermoso JA (2005) Probing the determinants of substrate specificity of a feruloyl esterase, AnFaeA, from Aspergillus niger. FEBS J 272:4362–4371

Filho EX, Puls J, Coughlan MP (1996) Purification and characterization of two arabinofuranosidases from solid-state cultures of the fungus Penicillium capsulatum. Appl Environ Microbiol 62:168–173

Foreman PK, Brown D, Dankmeyer L, Dean R, Diener S, Dunn-Coleman NS, Goedegebuur F, Houfek TD, England GJ, Kelley AS (2003) Transcriptional regulation of biomass-degrading enzymes in the filamentous fungus Trichoderma reesei. J Biol Chem 278:31988–31997

Galagan JE, Calvo SE, Cuomo C, Ma L-J, Wortman JR, Batzoglou S, Lee S-I, Basturkmen M, Spevak CC, Clutterbuck J, Kapitonov V, Jurka J, Scazzocchio C, Farman M, Butler J, Purcell S, Harris S, Braus GH, Draht O, Busch S, D'enfert C, Bouchier C, Goldman GH, Bell-Pedersen D, Griffiths-Jones S, Doonan JH, Yu J, Vienken K, Pain A, Freitag M, Selker EU, Archer DB, Penalva MA, Oakley BR, Momany M, Tanaka T, Kumagai T, Asai K, Machida M, Nierman WC, Denning DW, Caddick M, Hynes M, Paoletti M, Fischer R, Miller B, Dyer P, Sachs MS, Osmani SA, Birren BW (2005) Sequencing of Aspergillus nidulans and comparative analysis with A. fumigatus and A. oryzae. Nature 438:1105–1115

Garber M, Grabherr MG, Guttman M, Trapnell C (2011) Computational methods for transcriptome annotation and quantification using RNA-seq. Nat Methods 8:469–477

Gibbons JG, Rokas A (2013) The function and evolution of the Aspergillus genome. Trends Microbiol 21:14–22

Goffeau A, Barrell B, Bussey H, Davis R, Dujon B, Feldmann H, Galibert F, Hoheisel J, Jacq C, Johnston M (1996) Life with 6000 genes. Science 274:546–567

Goldbeck R, Damásio AR, Gonçalves TA, Machado CB, Paixão DA, Wolf LD, Mandelli F, Rocha GJ, Ruller R, Squina FM (2014) Development of hemicellulolytic enzyme mixtures for plant biomass deconstruction on target biotechnological applications. Appl Microbiol Biotechnol 98:8513–8525

Goldemberg J (2007) Ethanol for a sustainable energy future. Science 315:808–810

Gonçalves TA, Damásio ARL, Segato F, Alvarez TM, Bragatto J, Brenelli LB, Citadini APS, Murakami MT, Ruller R, Paes Leme AF, Prade RA, Squina FM (2012) Functional characterization and synergic action of fungal xylanase and arabinofuranosidase for production of xylooligosaccharides. Bioresour Technol 119:293–299

Gottschalk LMF, Oliveira RA, da Silva Bon EP (2010) Cellulases, xylanases, β-glucosidase and ferulic acid esterase produced by Trichoderma and Aspergillus act synergistically in the hydrolysis of sugarcane bagasse. Biochem Eng J 51:72–78

Häkkinen M, Arvas M, Oja M, Aro N, Penttilä M, Saloheimo M, Pakula TM (2012) Re-annotation of the CAZy genes of Trichoderma reesei and transcription in the presence of lignocellulosic substrates. Microb Cell Fact 11:134

Henrissat B (1991) A classification of glycosyl hydrolases based on amino acid sequence similarities. Biochem J 280:309–316

Henrissat B, Bairoch A (1993) New families in the classification of glycosyl hydrolases based on amino acid sequence similarities. Biochem J 293:781–788

Hermoso JA, Sanz-Aparicio J, Molina R, Juge N, González R, Faulds CB (2004) The crystal structure of feruloyl esterase A from Aspergillus niger suggests evolutive functional convergence in feruloyl esterase family. J Mol Biol 338:495–506

Horta MAC, Vicentini R, da Silva Delabona P, Laborda P, Crucello A, Freitas S, Kuroshu RM, Polikarpov I, da Cruz Pradella JG, AP S (2014) Transcriptome profile of Trichoderma harzianum IOC-3844 induced by sugarcane bagasse. PLoS One 9:e88689

Jacobsen SE, Wyman CE (2002) Xylose monomer and oligomer yields for uncatalyzed hydrolysis of sugarcane bagasse hemicellulose at varying solids concentration. Ind Eng Chem Res 41:1454–1461

Khaldi N, Wolfe KH (2008) Elusive origins of the extra genes in Aspergillus oryzae. PLoS One 3:e3036

Kim TJ (2008) Microbial exo- and endo-arabinosyl hydrolases: structure, function, and application in L-arabinose production. In: Carbohydrate-active enzymes. Woodhead Publishing Ltd., Cambridge, pp 229–257

Kroon PA, Williamson G (1996) Purification and characterization of a novel esterase induced by growth of Aspergillus niger on sugar beet pulp. Biotechnol Appl Biochem 23:255–262

Kroon PA, Williamson G, Fish NM, Archer DB, Belshaw NJ (2000) A modular esterase from Penicillium funiculosum which releases ferulic acid from plant cell walls and binds crystalline cellulose contains a carbohydrate binding module. Eur J Biochem 267:6740–6752

Kubicek CP, Herrera-Estrella A, Seidl-Seiboth V, Martinez DA, Druzhinina IS, Thon M, Zeilinger S, Casas-Flores S, Horwitz BA, Mukherjee PK (2011) Comparative genome sequence analysis underscores mycoparasitism as the ancestral life style of Trichoderma. Genome Biol 12:R40

Lavarack BP, Griffin GJ, Rodman D (2002) The acid hydrolysis of sugarcane bagasse hemicellulose to produce xylose, arabinose, glucose and other products. Biomass Bioenergy 23: 367–380

Lee RC, Hrmova M, Burton RA, Lahnstein J, Fincher GB (2003) Bifunctional family 3 glycoside hydrolases from barley with α-L-arabinofuranosidase and β-D-xylosidase activity: characterization, primary structures and COOH-terminal processing. J Biol Chem 278:5377–5387

Levasseur A, Navarro D, Punt PJ, Belaïch J-P, Asther M, Record E (2005) Construction of engineered bifunctional enzymes and their overproduction in Aspergillus niger for improved enzymatic tools to degrade agricultural by-products. Appl Environ Microbiol 71:8132–8140

Levasseur A, Saloheimo M, Navarro D, Andberg M, Monot F, Nakari-Setala T, Asther M, Record E (2006) Production of a chimeric enzyme tool associating the Trichoderma reesei swollenin with the Aspergillus niger feruloyl esterase A for release of ferulic acid. Appl Microbiol Biotechnol 73:872–880

Lubertozzi D, Keasling JD (2009) Developing Aspergillus as a host for heterologous expression. Biotechnol Adv 27:53–75

Macdonald J, Doering M, Canam T, Gong Y, Guttman D, Campbell M, Master E (2011) Transcriptomic responses of the softwood-degrading white-rot fungus Phanerochaete carnosa during growth on coniferous and deciduous wood. Appl Environ Microbiol 77:3211–3218

Macedo I, Seabra J, da Silva J (2008) Greenhouse gases emissions in the production and use of ethanol from sugarcane in Brazil: The 2005/2006 averages and a prediction for 2020. Biomass Bioenergy 32:582–595

Machida M, Asai K, Sano M, Tanaka T, Kumagai T, Terai G, Kusumoto K-I, Arima T, Akita O, Kashiwagi Y (2005) Genome sequencing and analysis of Aspergillus oryzae. Nature 438: 1157–1161

Mackenzie CR, Bilous D (1988) Ferulic acid esterase activity from Schizophyllum commune. Appl Environ Microbiol 54:1170–1173

Maeda H, Sano M, Maruyama Y, Tanno T, Akao T, Totsuka Y, Endo M, Sakurada R, Yamagata Y, Machida M, Akita O, Hasegawa F, Abe K, Gomi K, Nakajima T, Iguchi Y (2004) Transcriptional analysis of genes for energy catabolism and hydrolytic enzymes in the filamentous fungus Aspergillus oryzae using cDNA microarrays and expressed sequence tags. Appl Microbiol Biotechnol 65:74–83

Maehara T, Fujimoto Z, Ichinose H, Michikawa M, Harazono K, Kaneko S (2014) Crystal structure and characterization of the glycoside hydrolase family 62 α-l-arabinofuranosidase from Streptomyces coelicolor. J Biol Chem 289:7962–7972

Mai V, Wiegel J, Lorenz WW (2000) Cloning, sequencing, and characterization of the bifunctional xylosidase–arabinosidase from the anaerobic thermophile Thermoanaerobacter ethanolicus. Gene 247:137–143

Mallet LV, Becq J, Deschavanne P (2010) Whole genome evaluation of horizontal transfers in the pathogenic fungus Aspergillus fumigatus. BMC Genomics 11:171

Mardis ER (2008) The impact of next-generation sequencing technology on genetics. Trends Genet 24:133–141

Martinez D, Berka RM, Henrissat B, Saloheimo M, Arvas M, Baker SE, Chapman J, Chertkov O, Coutinho PM, Cullen D, Danchin EGJ, Grigoriev IV, Harris P, Jackson M, Kubicek CP, Han CS, Ho I, Larrondo LF, de Leon AL, Magnuson JK, Merino S, Misra M, Nelson B, Putnam N, Robbertse B, Salamov AA, Schmoll M, Terry A, Thayer N, Westerholm-Parvinen A, Schoch

CL, Yao J, Barabote R, Nelson MA, Detter C, Bruce D, Kuske CR, Xie G, Richardson P, Rokhsar DS, Lucas SM, Rubin EM, Dunn-Coleman N, Ward M, Brettin TS (2008) Genome sequencing and analysis of the biomass-degrading fungus Trichoderma reesei (syn. Hypocrea jecorina). Nat Biotechnol 26:553–560

Mcauley KE, Svendsen A, Patkar SA, Wilson KS (2004) Structure of a feruloyl esterase from Aspergillus niger. Acta Crystallogr Sect D 60:878–887

Moreira LRS, Ferreira GV, Santos SST, Ribeiro APS, Siqueira FG, Ferrei Rafilho EX (2012) The hydrolysis of agro-industrial residues by holocellulose-degrading enzymes. Braz J Microbiol 43:498–505

Moreira LRS, Milanezi NVG, Filho EXF (2011) Enzymology of plant cell wall breakdown: an update. In: Buckeridge MS, Goldman GH (eds) Routes to cellulosic ethanol. Springer, New York

NCBI (2015) National Center for Biotechnology Information—genome [online]. Accessed 23 Mar 2015

Nierman WC, Pain A, Anderson MJ, Wortman JR, Kim HS, Arroyo J, Berriman M, Abe K, Archer DB, Bermejo C, Bennett J, Bowyer P, Chen D, Collins M, Coulsen R, Davies R, Dyer PS, Farman M, Fedorova N, Fedorova N, Feldblyum TV, Fischer R, Fosker N, Fraser A, Garcia JL, Garcia MJ, Goble A, Goldman GH, Gomi K, Griffith-Jones S, Gwilliam R, Haas B, Haas H, Harris D, Horiuchi H, Huang J, Humphray S, Jimenez J, Keller N, Khouri H, Kitamoto K, Kobayashi T, Konzack S, Kulkarni R, Kumagai T, Lafton A, Latge J-P, Li W, Lord A, Lu C, Majoros WH, May GS, Miller BL, Mohamoud Y, Molina M, Monod M, Mouyna I, Mulligan S, Murphy L, O'neil S, Paulsen I, Penalva MA, Pertea M, Price C, Pritchard BL, Quail MA, Rabbinowitsch E, Rawlins N, Rajandream M-A, Reichard U, Renauld H, Robson GD, de Cordoba SR, Rodriguez-Pena JM, Ronning CM, Rutter S, Salzberg SL, Sanchez M, Sanchez-Ferrero JC, Saunders D, Seeger K, Squares R, Squares S, Takeuchi M, Tekaia F, Turner G, de Aldana CRV, Weidman J, White O, Woodward J, Yu J-H, Fraser C, Galagan JE, Asai K, Machida M, Hall N, Barrell B, Denning DW (2005) Genomic sequence of the pathogenic and allergenic filamentous fungus Aspergillus fumigatus. Nature 438:1151–1156

Pandey A, Soccol CR, Nigam P, Soccol VT (2000) Biotechnological potential of agro-industrial residues. I: sugarcane bagasse. Bioresour Technol 74:69–80

Payne GA, Nierman WC, Wortman JR, Pritchard BL, Brown D, Dean RA, Bhatnagar D, Cleveland TE, Machida M, Yu J (2006) Whole genome comparison of Aspergillus flavus and A. oryzae. Med Mycol 44:9–11

Pel HJ, de Winde JH, Archer DB, Dyer PS, Hofmann G, Schaap PJ, Turner G, de Vries RP, Albang R, Albermann K (2007) Genome sequencing and analysis of the versatile cell factory Aspergillus niger CBS 513.88. Nat Biotechnol 25:221–231

Peng F, Ren J-L, Xu F, Bian J, Peng P, Sun R-C (2009) Comparative study of hemicelluloses obtained by graded ethanol precipitation from sugarcane bagasse. J Agric Food Chem 57:6305–6317

Portnoy T, Margeot A, Seidl-Seiboth V, Le Crom S, Chaabane FB, Linke R, Seiboth B, Kubicek CP (2011) Differential regulation of the cellulase transcription factors XYR1, ACE2, and ACE1 in Trichoderma reesei strains producing high and low levels of cellulase. Eukaryot Cell 10:262–271

Prates JAM, Tarbouriech N, Charnock SJ, Fontes CMGA, Ferreira LSMA, Davies GJ (2001) The structure of the feruloyl esterase module of xylanase 10B from Clostridium thermocellum provides insights into substrate recognition. Structure 9:1183–1190

Ralph J, Lundquist K, Brunow G, Lu F, Kim H, Schatz P, Marita J, Hatfield R, Ralph S, Christensen J, Boerjan W (2004) Lignins: natural polymers from oxidative coupling of 4-hydroxyphenylpropanoids. Phytochem Rev 3:29–60

Ries L, Pullan ST, Delmas S, Malla S, Blythe MJ, Archer DB (2013) Genome-wide transcriptional response of Trichoderma reesei to lignocellulose using RNA sequencing and comparison with Aspergillus niger. BMC Genomics 14:541

Rye CS, Withers SG (2000) Glycosidase mechanisms. Curr Opin Chem Biol 4:573–580

Saha BC (2000) α-l-Arabinofuranosidases: biochemistry, molecular biology and application in biotechnology. Biotechnol Adv 18:403–423

Semova N, Storms R, John T, Gaudet P, Ulycznyj P, Min XJ, Sun J, Butler G, Tsang A (2006) Generation, annotation, and analysis of an extensive Aspergillus niger EST collection. BMC Microbiol 6:7

Shallom D, Belakhov V, Solomon D, Gilead-Gropper S, Baasov T, Shoham G, Shoham Y (2002) The identification of the acid–base catalyst of α-arabinofuranosidase from Geobacillus stearothermophilus T-6, a family 51 glycoside hydrolase. FEBS Lett 514:163–167

Shallom D, Leon M, Bravman T, Ben-David A, Zaide G, Belakhov V, Shoham G, Schomburg D, Baasov T, Shoham Y (2005) Biochemical characterization and identification of the catalytic residues of a family 43 β-D-xylosidase from Geobacillus stearothermophilus T-6. Biochemistry 44:387–397

Shofiqur Rahman AKM, Kato K, Kawai S, Takamizawa K (2003) Substrate specificity of the α-l-arabinofuranosidase from Rhizomucor pusillus HHT-1. Carbohydr Res 338:1469–1476

Siqueira FG, Filho EXF (2010) Plant cell wall as a substrate for the production of enzymes with industrial applications. Mini-Rev Org Chem 7:55–60 Bentham Science Publishers

Sørensen A, Teller PJ, Lübeck PS, Ahring BK (2011) Onsite enzyme production during bioethanol production from biomass: screening for suitable fungal strains. Appl Biochem Biotechnol 164:1058–1070

Souza WR, de Gouvea PF, Savoldi M, Malavazi I, de Souza Bernardes LA, Goldman MHS, de Vries RP, de Castro Oliveira JV, Goldman GH (2011) Transcriptome analysis of Aspergillus niger grown on sugarcane bagasse. Biotechnol Biofuels 4:1–17

Souza WR, Maitan-Alfenas GP, de Gouvêa PF, Brown NA, Savoldi M, Battaglia E, Goldman MHS, de Vries RP, Goldman GH (2013) The influence of Aspergillus niger transcription factors AraR and XlnR in the gene expression during growth in d-xylose, l-arabinose and steam-exploded sugarcane bagasse. Fungal Genet Biol 60:29–45

Srivastava M, Shahid M, Pandey S, Singh A, Kumar V (2014) Trichoderma genome to genomics: a review. J Data Min Genomics Proteomics 5(3):162

Sun J-X, Sun R, Sun X-F, Su Y (2004a) Fractional and physico-chemical characterization of hemicelluloses from ultrasonic irradiated sugarcane bagasse. Carbohydr Res 339:291–300

Sun JX, Sun XF, Sun RC, Su YQ (2004b) Fractional extraction and structural characterization of sugarcane bagasse hemicelluloses. Carbohydr Polym 56:195–204

Sun X-F, Wang H, Zhang G, Fowler P, Rajaratnam M (2011) Extraction and characterization of lignins from maize stem and sugarcane bagasse. J Appl Polym Sci 120:3587–3595

Szczerbowski D, Pitarelo AP, Zandoná Filho A, Ramos LP (2014) Sugarcane biomass for biorefineries: comparative composition of carbohydrate and non-carbohydrate components of bagasse and straw. Carbohydr Polym 114:95–101

Tabka MG, Herpoël-Gimbert I, Monod F, Asther M, Sigoillot JC (2006) Enzymatic saccharification of wheat straw for bioethanol production by a combined cellulase xylanase and feruloyl esterase treatment. Enzyme Microb Technol 39:897–902

Tateishi A, Kamiyoshihara Y, Matsuno J, Miyohashi F, Shiba H, Kanayama Y, Watanabe K, Nomura K, Inoue H (2014) Heterologous expression of tomato glycoside hydrolase family 3 α-L-arabinofuranosidase/β-xylosidases in tobacco suspension cultured cells and synergic action of a family 51 isozyme under antisense suppression of the enzyme. Physiol Plant 150:238–251

Topakas E, Vafiadi C, Christakopoulos P (2007) Microbial production, characterization and applications of feruloyl esterases. Process Biochem 42:497–509

Udatha DBRKG, Mapelli V, Panagiotou G, Olsson L (2012) Common and distant structural characteristics of feruloyl esterase families from Aspergillus oryzae. PLoS One 7:e39473

Vafiadi C, Topakas E, Christakopoulos P, Faulds CB (2006) The feruloyl esterase system of Talaromyces stipitatus: determining the hydrolytic and synthetic specificity of TsFaeC. J Biotechnol 125:210–221

van Dyk JS, Pletschke BI (2012) A review of lignocellulose bioconversion using enzymatic hydrolysis and synergistic cooperation between enzymes—factors affecting enzymes, conversion and synergy. Biotechnol Adv 30:1458–1480

Varnai A, Costa T, Faulds C, Milagres A, Siika-Aho M, Ferraz A (2014) Effects of enzymatic removal of plant cell wall acylation (acetylation, p-coumaroylation, and feruloylation) on accessibility of cellulose and xylan in natural (non-pretreated) sugar cane fractions. Biotechnol Biofuels 7:153

Wang T-Y, Chen H-L, Lu M, Chen Y-C, Sung H-M, Mao C-T, Cho H-Y, Ke H-M, Hwa T-Y, Ruan S-K (2011) Functional characterization of cellulases identified from the cow rumen fungus Neocallimastix patriciarum W5 by transcriptomic and secretomic analyses. Biotechnol Biofuels 4:24

Wong DS (2006) Feruloyl esterase: a key enzyme in biomass degradation. Appl Biochem Biotechnol 133:87–112

Xu F, Sun JX, Liu CF, Sun RC (2006) Comparative study of alkali- and acidic organic solvent-soluble hemicellulosic polysaccharides from sugarcane bagasse. Carbohydr Res 341:253–261

Xu F, Sun R-C, Sun J-X, Liu C-F, He B-H, Fan J-S (2005) Determination of cell wall ferulic and p-coumaric acids in sugarcane bagasse. Anal Chim Acta 552:207–217

Zeng Y, Yin X, Wu M-C, Yu T, Feng F, Zhu T-D, Pang Q-F (2014) Expression of a novel feruloyl esterase from Aspergillus oryzae in Pichia pastoris with esterification activity. J Mol Catal B: Enzym 110:140–146

Znameroski EA, Coradetti ST, Roche CM, Tsai JC, Iavarone AT, Cate JH, Glass NL (2012) Induction of lignocellulose-degrading enzymes in Neurospora crassa by cellodextrins. Proc Natl Acad Sci 109:6012–6017

Chapter 7
Applied Metagenomics for Biofuel Development and Environmental Sustainability

Antônio Djalma Nunes Ferraz Júnior, André Ricardo L. Damásio, Douglas Antonio Alvaredo Paixão, Thabata Maria Alvarez, and Fabio Marcio Squina

Abstract Traditional ethanol from sugarcane, also known as first generation ethanol (1G), is one of the greatest technological achievements in the modern history of Brazil. Second-generation bioethanol (2G) from sugarcane bagasse is considered a rational and sustainable alternative to the expansion of the sugarcane industry. However, 2G technology is not yet as robust as 1G ethanol. In this chapter, metagenomics approaches are contextualized within the bioethanol production chain. A comprehensive description is presented on novel carbohydrate-active enzymes (CAZymes) discovered from uncultivated microorganisms with potential utilization in biomass conversion strategies. In addition, this chapter also focuses on the potential of the development of microbial communities to convert residual streams derived from ethanol into valuable products such as biogas and biofertilizer. Finally, a case study is presented that describes the application of large-scale sequencing to assess microbial communities from anaerobic reactors involved in hydrogen production.

Keywords Biofuels • Metagenomics • Enzymes • Anaerobic digestion

---

A.D.N. Ferraz Júnior • D.A.A. Paixão • T.M. Alvarez (✉) • F.M. Squina (✉)
Laboratório Nacional de Ciência e Tecnologia do Bioetanol (CTBE),
Centro Nacional de Pesquisa em Energia e Materiais (CNPEM),
Caixa Postal 6192, CEP 13083-970, Campinas, São Paulo, Brazil

Professional Master in Industrial Biotechnology,
Positivo University, R. Prof. Pedro Viriato Parigot de Souza, 5300, Cidade Industrial,
81280-330, Campinas, Paraná, Brazil
e-mail: thabata.alvarez@up.edu.br; fabio.squina@bioetanol.org.br

A.R.L. Damásio
Laboratório Nacional de Ciência e Tecnologia do Bioetanol (CTBE),
Centro Nacional de Pesquisa em Energia e Materiais (CNPEM),
Caixa Postal 6192, CEP 13083-970, Campinas, São Paulo, Brazil

Department of Biochemistry and Tissue Biology, Institute of Biology,
University of Campinas, Rua Monteiro Lobato 255, Cidade Universitária
Zeferino Vaz, 13083-862 Campinas, São Paulo, Brazil

M.S. Buckeridge, A.P. De Souza (eds.), *Advances of Basic Science for Second Generation Bioethanol from Sugarcane*, DOI 10.1007/978-3-319-49826-3_7

7.1 Introduction

The issues associated with the use of fossil fuels observed in the last decades have resulted in the development of alternative sources to meet the increasing demand for energy, especially in the automotive sector. The biofuel industry and especially ethanol production can be highlighted among these alternative sources.

Traditional ethanol from sugarcane, also known as first generation ethanol (1G), is one of the greatest technological achievements in the modern history of Brazil (De Souza et al. 2014). The innovation started in 1975 with the Pro-Alcohol Program, which was responsible for reducing the country's dependence on fossil fuels (Goldemberg et al. 2004). Along the years, several factors influenced the development of ethanol industry, such as political transition and economic instability. However, in the early 2000s, the development of flex fuel vehicles and the rise of oil prices renewed the interest for the use of ethanol as a biofuel (Nass et al. 2007).

At the end of the 2013/2014 Brazilian sugarcane harvesting season, approximately 27 billion liters of ethanol was produced. According to estimates of the Brazilian development bank (BNDES), the accumulated deficit of ethanol supply in the period of 2011–2015 is 32.4 billion liters. Thus, it is necessary to maximize the production of biofuels in the coming years (Milanez et al. 2012).

Cellulosic ethanol, or the second generation of ethanol production (2G ethanol), is seen as an alternative to the segment expansion that can facilitate increased total ethanol production by up to 50% (Raízen 2015). The source of 2G ethanol is mainly the sugarcane milling bagasse generated from 1G ethanol production, indicating no need to expand the planted area and, thus, eliminating competition with the agricultural production of food. Furthermore, the technology of 2G ethanol may be designed to allow full integration with the existing structure in the plants, thus optimizing the installation and operating costs (Fig. 7.1).

The 2G ethanol is now a reality in Brazil. In 2014, the first two industrial plants for 2G ethanol production in full scale started its operation in the country (GranBio 2015; Raízen 2015). However, 2G ethanol technology is not as robust as 1G ethanol and must be improved in two specific and crucial steps: pretreatment and hydrolysis of the lignocellulosic material (see also Chaps. 5 and 6). In addition, there is a clear demand for the development of sustainable processes focusing on the conversion of residual streams from ethanol production into valuable products.

In this chapter, we highlight how metagenomics can be inserted within the context of an integrated biorefinery (1G2G) based on sugarcane (Fig. 7.1), to boost bioethanol competitiveness against fossil fuels. In this sense, firstly, we are going to provide an overview of metagenomics and the main landmark achievements (Sect. 7.2), followed by the focus on the application of metagenomics for improvement of enzymatic hydrolysis step through the identification of novel enzymes and biocatalysts (Sect. 7.3) and on the advances of applied engineering research promoted by metagenomics (Sect. 7.4), which are outlined in a case study (Sect. 7.4.1).

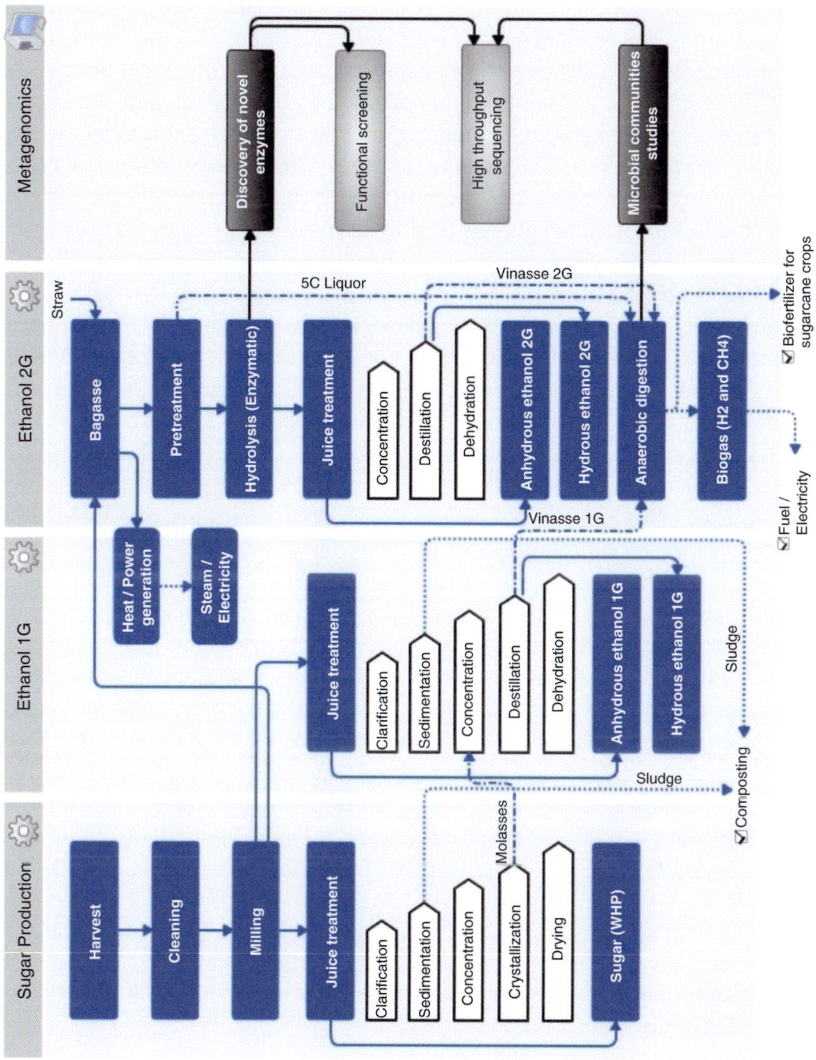

Fig. 7.1 Flowchart of an integrated biorefinery (1G2G) based on sugar cane and insertion of metagenomics for biofuel development

7.2 Brief Overview of Metagenomics Approaches

The term "metagenome" was created by Handelsman and colleagues and is used to describe the collection of genomic DNA extracted from a microbial community (Handelsman et al. 1998).

During the development of microbiology, evidence accumulated supporting the idea that cultured microorganisms did not fully represent the microbial world (Handelsman 2004). In 1985, Staley and Konopka used the term "great plate count anomaly" to refer to the inequality observed when comparing the number of cells counted by dilution plating with the number of cells directly visualized by microscopy (Staley and Konopka 1985; Amann et al. 1995; Handelsman 2004). The explanation for this was related to the fact that most of the cells counted by the microscope were not able to form visible colonies, mainly because of the restricted cultivation conditions applied or because some microorganisms can display a nonculturable state (Amann et al. 1995).

The direct analysis of small subunit ribosomal RNAs enabled a description of microbial diversity that bypassed the need for cultivation. Initially, these studies relied on the direct isolation of 5S rRNA by high resolution gel electrophoresis or by shotgun cloning of 16S rRNA followed by sequencing analysis (Pace et al. 1986). An important advancement in this area was the development of PCR (Polymerase Chain Reaction) technology (Saiki et al. 1988) along with the selective amplification of ribosomal RNAs coding sequences, confirming that uncultured microorganisms represent the majority of microbial diversity present in environmental samples (Amann et al. 1995; Handelsman 2004). It is estimated that culturable bacteria can represent from 0.001 to 0.3% of the total cell counts, depending on the type of environment (Amann et al. 1995).

One of the most important steps common to all metagenomic studies is the choice of an appropriate protocol for metagenomic DNA extraction because it will determine the purity and integrity of the metagenomic DNA. The extraction methods are divided into two types: direct and indirect. The direct methods were first reported by Ogram et al. (1987) and, in general, consist of performing the cell lysis in situ followed by removal and purification of metagenomic DNA. In contrast, the indirect methods, first described by Holben et al. (1988), require the separation of microbial cells prior to the lysis step (Bertrand et al. 2005; Daniel 2005). By comparing the application of both methods in the extraction of metagenomic DNA from a variety of samples, it was observed that direct methods are generally related to the greatest DNA yield, while indirect methods resulted in DNA samples of higher purity (Steffan et al. 1988; Leff et al. 1995; Gabor et al. 2003).

After obtaining metagenomic DNA with the intent to study the taxonomic composition of a given sample or for the identification of metabolic pathways and genes of interest, two main approaches are conducted: functional- and sequence-driven approaches.

7.2.1 Functional Metagenomic Approaches

Function-driven approaches require the fragmentation and cloning of metagenomic DNA into vectors such as plasmids, fosmids, and cosmids and transformation of a host cell, generating the so-called metagenomic library, followed by the screening of those clones that are able to express a desirable metabolic function (Schloss and Handelsman 2003). The main advantage of this approach is the possibility of finding completely new genes. On the other hand, an important disadvantage is that a large number of clones must be analyzed to find a positive match, because the technique depends on the compatibility between the cloned gene and the expression system from the host cell. Furthermore, in most cases, the secretion of gene products into the extracellular medium, insertion of full-length genes, start codon recognition by the host strain, and post-translational modifications must be considered (Handelsman 2004; Li et al. 2009). In addition, specific natural conditions of the target microbial community, including low population density, can also be a limiting factor to access the genetic material of these species. Several studies have shown that the number of clones screened in a metagenomic library can reach 37,000 or more (Rees et al. 2003; Grant et al. 2004; Feng et al. 2007).

The low frequency of positive clones found in metagenomic DNA libraries has been a recurring problem for researchers screening genes based on function-driven approaches. Several strategies for targeting gene enrichment have been successfully suggested as an alternative to provide a larger number of positive clones in a metagenomic DNA library (Entcheva et al. 2001; Knietsch et al. 2003; Gabor et al. 2004; Grant et al. 2004; Daniel 2005).

The development of targeted microbial enrichment protocols could be one way to modulate the abundance of microorganisms that are at relatively low frequency, as well to provide increased number of genes and pathways for a certain scientific application (Brenner et al. 2008; Pandhal and Noirel 2014).

New approaches to obtain microbial enrichment have shown great potential for biotechnological applications. Basically, these enrichments are defined as a mix of species from microbial communities, bacteria and fungi, which, in synergy, can perform specific metabolic functions that represent a challenge for single microbial strains (Brenner et al. 2008). Selective pressure on the microbial communities through nutritional conditions such as the carbon and nitrogen source and even physical conditions such as temperature allow the selection of metabolically specialized communities. Gabor et al. (2004) showed that metagenomic DNA libraries from enriched microbial communities grown in several sources of nitrogen had better performance to prospect amidases genes when compared with nonenrichment DNA libraries. This strategy results not only in a reduction of costs but also in time spent to screen metagenomic DNA libraries. In this context, highly selective microbial communities, as described above, can be potential genetic sources for exploration and discovery of new plant cell wall converting enzymes.

7.2.2 Sequence-Driven Metagenomic Approaches

In general, sequence-driven approaches involve the use of PCR primers and hybridization probes with designs based on conserved regions of known genes to amplify or identify genes of interest directly in metagenomic DNA or metagenomic libraries (Schloss and Handelsman 2003; Simon and Daniel 2011). The main drawback is that this approach discloses genes that are related to already-known protein families available in public databases (Daniel 2005; Simon and Daniel 2011).

Among sequence-driven approaches, the application of high-throughput sequencing of genetic material derived from environmental samples has become a widespread practice. A landmark paper was published in 2004 by Tyson that applied large-scale shotgun sequencing of DNA samples extracted from a low complexity microbial community from a natural acidophilic biofilm. In this case, shotgun sequencing resulted in 76.2 Mbp (mega base pairs), which enabled the almost complete reconstruction of two genomes (Leptospirillum group II and Ferroplasma type II) and partial reconstruction of another three. Moreover, it provided information regarding the establishment of relationships of the metabolic pathways among the predominant species identified and also related to the mechanisms applied for carbon and nitrogen fixation and energy generation (Tyson et al. 2004).

Another important paper in this area was also published in 2004, by Venter, which described the application of whole-genome shotgun sequencing for the study of microbial populations from seawater samples collected from the Sargasso Sea. In total, more than 1.6 Gbp (giga base pairs) were generated from this work, helping to shed light on the microbial diversity of this particular marine environment, resulting in the identification of 148 previously unknown phylotypes (97% sequence similarity cutoff) and the prediction of more than one million genes related to a variety of metabolic pathways such as amino acid biosynthesis, energy metabolism, and protein synthesis (Venter et al. 2004).

In 2005, Tringe and coworkers performed a comprehensive analysis to investigate the usefulness of whole-genome shotgun sequencing in the case of complex microbial communities from agricultural soil and deep sea whale skeletons, despite the low assembly rate due to the intrinsic heterogeneity of the sequences generated. One of the major contributions of this work was to show that, even with the unassembled data, it was possible to observe predicted genes that act as specific fingerprints for a particular habitat using a gene-centric approach. Additionally, a correlation was suggested to exist between the metabolic challenges imposed by different habitats and the composition of predicted genes from different microbial communities (Tringe et al. 2005; Tringe and Rubin 2005).

The three papers cited above applied the so-called first generation automated sequencing technologies that were based on Sanger's sequencing method (Sanger et al. 1977). Although "First Generation Sequencing Technologies" enabled important achievements, including the sequencing of human genome, the demand for cheaper technologies with higher throughputs in a reduced period of time stimulated the development of Next Generation Sequencing Technologies (NGS) (Grada and Weinbrecht 2013; Barba et al. 2014). A variety of sequencing platforms that apply different biochemistry principles falls in the NGS category, such as 454 pyrosequencing from Roche, Solexa GA from Illumina, SOLiD from Applied Biosystems, and Heliscope from Helicos (MacLean et al. 2009). The common features among NGS technologies are related to the massive parallel sequencing throughput and the ability to circumvent traditional techniques required for sample preparation in Sanger's sequencing such as vector-based cloning, since library preparation for NGS involves basic DNA fragmentation and ligation to specific adaptor sequences (Mardis 2008). In addition to these advantages, NGS enabled a dramatic drop in sequencing cost, although it requires considerable investments for the establishment and maintenance of sequencing facilities (El-Metwally et al. 2014). For many years, the main drawback of NGS was related to the length of sequence reads, which used to be substantially smaller in comparison to Sanger's reads (Dröge and McHardy 2012; Mardis 2008). However, several technological developments have been made to increase the read length, such as Illumina HiSeq 2500 that produces reads of up to 250 bp (base pairs) paired-end, GS FLX Titanium XL+ with reads of 700 bp, and Pacific Biosciences RSII that is able to generate at least half of the reads higher than 10 kbp (kilo base pairs) (Buermans and den Dunnen 2014).

In comparison to traditional approaches, the application of next-generation sequencing to metagenomics studies has significantly increased the sampling depth (Simon and Daniel 2011). In 2011, Hess et al. performed a deep study about the microbial community and genes involved in biomass degradation in cow rumen by sequencing the metagenomic DNA extracted from the microorganisms adherent to plant fibers present in nylon bags that were incubated directly in the rumen. In an attempt to facilitate genome assembly, libraries of 200 bp, 3 kbp, and 5 kbp were sequenced on Illumina GAIIx, and a library of an average insert size of 300 bp was sequenced on Illumina HiSeq2000, generating in total 268 Gbp of sequence data. After assembly and prediction of open reading frames, 27,755 candidate genes exhibiting significant homology to functional domains of catalytic domains or carbohydrate-binding modules from carbohydrate-active enzymes were identified. By comparing the amino acid sequences from this total number of candidate genes against the NCBI-nr (NCBI nonredundant) database, it was possible to observe that the majority of these genes displayed less than 75% sequence identity to deposited genes, while the comparison with NCBI-env (environmental DNA) indicated that most of these genes have not been identified by previous works (Hess et al. 2011).

Also in 2011, Mackelprang and coworkers described the use of the Illumina GAII platform for the sequencing of metagenomic DNA extracted from frozen permafrost soil cores and their active layers before and during induced thaw, to evaluate the shifts in the microbial community and in their functional profiles, along with their relation to methane emissions. In total, 39.8 Gbp of raw sequence were generated, which enabled the reconstruction of the draft genome for new methanogen bacteria. In addition, metagenomic data provided insights about the cycling of carbon and nitrogen in permafrost soil samples exposed to thaw (Mackelprang et al. 2011).

In 2012, van der Lelie described a combination of Sanger sequencing and 454 pyrosequencing for the characterization of a microbial consortium involved in the conversion of poplar wood chips under anaerobic conditions. As a result, 675 Mbp sequencing data were generated and, after quality filter and exclusion of nonredundant reads, 44,600 contigs were generated in addition to the 1.24 M singletons. The analysis of these data enabled the identification of candidate genes with homology to a wide variety of carbohydrate-active enzymes (Lombard et al. 2014) and also to some fungal oxidase families (according to the FOLy database, Levasseur et al. 2008), shedding light on the mechanism of conversion of poplar hardwood in this enriched microbial community (van der Lelie et al. 2012).

A recent paper published by Mhuantong et al. (2015) described the construction and pyrosequencing of a fosmid library of DNA extracted from a microbial community naturally growing on sugarcane bagasse. As a result, 726,980 filtered reads were assembled into 17,829 contigs (average length of 1.84 kbp) and 185,543 nonredundant singletons (average length of 589 bp). The analysis of the data enabled the characterization of the taxonomic profile of the microbial community and the evaluation of genetic information related to genes coding for proteins involved in plant biomass conversion. Additionally, it was shown that the content of unique genes coding for proteins involved in biomass degradation were higher and similar to lignocellulolytic microbial communities in comparison with nonlignocellulolytic communities (Mhuantong et al. 2015).

In parallel with the advances in metagenomics approaches, new strategies are being developed for the culture and isolation of organisms from the so-called uncultured world. This is because the isolation of completely novel organisms and their genomes are able to provide a full picture of the metabolic pathways of novel organisms and, subsequently, result in novel enzymes and metabolites of interest. In this context, one example can be found in the paper published by Ling et al. (2015), which described the identification of a novel antibiotic termed "teixobactin" produced by a bacterium that was primarily grown in soil, which is its natural environment (Ling et al. 2015). Using the iChip technology (Nichols et al. 2010), individual colonies were screened for antibiotic activity followed by genome sequencing. As a result, the novel species temporarily named *Eleftheria terrae* was identified as being able to produce the new antibiotic teixobactin (Ling et al. 2015).

7.3 The Application of Metagenomics on Enzymatic Hydrolysis Processes

The enzymatic hydrolysis process has been considered advantageous due to its moderate reaction conditions, higher efficiency, the absence of substrate chemical modifications, and the absence of inhibitory compound release (Adsul et al. 2011). However, the enzymatic process is a sensitive process that requires the careful balancing of many variables such as (1) enzyme loading, (2) hydrolysis time, (3) solid content, and (4) enzyme cost. An optimal balancing of these four variables will be reflected in the cost and the yield of 2G ethanol.

Focusing on the enzymes that act on plant biomass conversion, the classes known to present significant importance in 2G ethanol production are as follows: glycoside hydrolases, polysaccharide lyases, carbohydrate esterases, lignin oxidases, and lignin-degrading auxiliary enzymes (Sweeney and Xu 2012). Nevertheless, a complete set of enzymes remains unknown (Himmel et al. 2007). In this sense, the application of metagenomic approaches may contribute to the discovery of carbohydrate-active enzymes (CAZymes) (Lombard et al. 2014).

First, before the description of metagenomic enzymes it is important to consider the structural characteristics of lignocellulosic plant biomass and the canonical view of the enzymatic activities involved in its degradation (see also Chaps. 2 and 3). For instance, sugarcane bagasse is mainly composed of cellulose (43%), hemicellulose (24%), and lignin (22%) (Khademi et al. 2002). Cellulose consists of D-glucose units linked by beta-1,4 glycosidic bonds, in a way that each glucose unit is positioned in a rotation of 180° from its adjacent molecule, resulting in long linear chains containing 2000–25,000 glucose residues (Segato et al. 2014). Each cellulose molecule is tightly bound to other molecules by multiple hydrogen bonds producing insoluble, rigid, and crystalline microfibrils (Carpita and McCann 2000, see Chap. 3). These cellulose fibers are prevented from collapsing into each other by hemicelluloses.

The canonical view of cellulose degradation requires at least three types of cellulases: cellobiohydrolases (CBHs), which attack the nonreducing or reducing ends of a cellulose chain producing cellobiose; endoglucanases, which cleave the internal linkages of cellulose molecules producing cellooligosaccharides; and β-glucosidases, which cleave cellobiose, or short cellooligosaccharides, into two free glucose molecules (Berghem et al. 1976, see Chap. 5). Recently, oxidative enzymes have been described as "enhancers" of cellulose enzymatic degradation, boosting the biomass conversion yield. This new class of enzymes includes the lytic polysaccharide monooxygenases (LPMOs) (Levasseur et al. 2013).

Hemicellulose is the second major polysaccharide fraction, and it is a complex component defined as polymers extractable by alkaline solutions. The hemicellulose fractions are named based on the main sugar in the polysaccharide backbone, such as xylan (xylose), arabinoxylan (xylose + arabinose), mannan (mannose), and xyloglucan (xylose + glucose) (Du Toit et al. 1984). The structural complexity of hemicellulose is directly related to the number of enzymes required for its complete degradation (De Souza et al. 2013, see Chap. 2). A detailed description of the enzymes involved in hemicellulose degradation can be found in a recent review (Segato et al. 2014, see also Chaps. 5 and 6).

The Carbohydrate-Active Enzymes database (Lombard et al. 2014) (CAZy; http://www.cazy.org) describes the families of structurally related catalytic and carbohydrate-binding modules of enzymes that degrade, modify, or create glycosidic bonds. A classification of glycoside hydrolases based on amino acid sequence similarities was published in 1991. A total of 291 sequences corresponding to 39 EC (Enzyme Commission) numbers were classified into 35 families (Henrissat 1991). The database, launched in 1999, has been continually updated; currently, the expansion of the enzymatic repertoire of the CAZy database has described 135 families of glycoside hydrolases (GH), 110 families of glycosyl transferases (GT), 24 families of polysaccharide lyases (PL), 16 families of carbohydrate esterases

Table 7.1 CAZymes isolated from metagenomes

Predicted function	CAZy family	Environment	Screening	Enzymatic properties			Substrate	Reference
				MW (kDa)	pH	*T* (°C)		
Xylanase	GH8	Adult lepidopteran	Functional	–	5.0	–	Azo-xylan	Brennan et al. (2004)
Xylanase	GH11	Adult isopteran	Functional	–	6.0	–	Azo-xylan	Brennan et al. (2004)
Amylase	–	Soil sample	Functional	58.0	9.0	42.0	Soluble starch	Yun et al. (2004)
Endo-1,4-beta-xylanase	GH8	Waste lagoon	Functional	45.9	6.0-7.0	20.0	Beech wood xylan	Lee et al. (2006)
Esterase (EstA3)	–	Drinking water from rubber-coated valves	Functional	42.0	9.0	50.0	*p*NP-acylesters; triglycerides; vinyl acids	Elend et al. (2006)
Esterase (EstCE1)	–	Soil sample	Functional	44.0	10.0	47.0	*p*NP-acylesters; triglycerides; vinyl acids	Elend et al. (2006)
Cellulase	GH5	Soil sample	Functional	42.1	6.5	45	CMC	Voget et al. (2006)
Endo-1,4-beta-xylanase	GH10	Soil sample	Functional	39.0	7.8	40.0	Birch wood xylan	Hu et al. (2008)
Endo-1,4-beta-glucanase	GH5	Buffalo rumen	Functional	62.0	4.5	45.0	CMC	Duan et al. (2009)
Endo-1,4-beta-glucanase	GH5	Antarctic soil	Functional	39.5	7.0	55.0	CMC	Berlemont et al. (2009)
Laccase	–	Marine sample	Sequence-based	47.8	7.5	45.0	Syringaldazine	Fang et al. (2011)
Pectinase	GH28	Soil sample	Sequence-based	47.9	7.0	70	Polygalacturonic acid	Singh et al. (2012)
Xylanase	GH10	Compost from pig manure and mushroom culture waste	Functional	39.9	7.0	40	Oat spelt xylan	Jeong et al. (2012)
Alpha-L-arabinofuranosidase	GH51	Fungus-growing termite	Functional	–	–	–	*p*NP-α-L-Ara*f*	Bastien et al. (2013)
Endo-1,4-beta-xylanase	GH11	Compost-soil	Functional	42.0	9.0	80.0	Birch wood xylan	Verma et al. (2013)
Endo-1,4-beta-xylanase	GH10	Sugarcane soil	Functional	40.0	6.0	45.0	Beech wood xylan	Alvarez et al. (2013a)

Endo-1,4-beta-glucanase	GH5	Sugarcane soil	Functional	34	7.0	50.0	CMC/beta-glucan	Alvarez et al. (2013b)
Beta-glucosidase (AS-Esc6)	GH3	Winter wheat soil	Functional	–	7.0-9.0	37.0	*p*NP-β-D-glucopyranosidase	Stroobants et al. (2014)
Beta-glucosidase (AS-Esc9)	GH1	Winter wheat soil	Functional	–	5.0-7.0	60.0	*p*NP-β-D-glucopyranosidase	Stroobants et al. (2014)
Endo-β-1,3-1,4-glucanase	GH8	Forest soil	Functional	41.7	6.0	50	Azo-barley-glucan	Kim et al. (2014)
Endo-1,4-beta-glucanase	GH5	Gut microflora of *Hermetia illucens*	Functional	40.0	7.0	50.0	–	Lee et al. (2014)
Endo-1,4-beta-glucanase		Soil sample	Functional	34.0	5.5	50.0	CMC	Xiang et al. (2014)
Endo-1,4-beta-glucanase		Seaweed microbiota	Functional	–	–	–	–	Martin et al. (2014)

(CE), and 13 families of auxiliary enzymes (AA) that cover redox enzymes that act in conjunction with GHs. Associated modules named carbohydrate-binding modules (CBM) are also classified into 81 families (Levasseur et al. 2013; Lombard et al. 2014).

The "classical" microbiological methods for enzyme bioprospection involve sampling from the target environment, cultivation/isolation of microorganisms, and global analysis of produced enzymes and other valuable products. These methods are time-consuming and require a laboratory structure to continuously maintain a library of microorganisms. However, bioprospection approaches based on metagenomic methods, described in Sect. 7.2, have contributed greatly to the expansion of the CAZy database.

The mining of environmental metagenomes has recently resulted in the identification of numerous new genes and corresponding putative CAZymes. Table 7.1 summarizes several cell wall-related GHs mined by function-based and sequence-based metagenomic approaches and some biochemical data of these enzymes. The low-throughput of function-based screening of environmental DNA clone libraries is the main "bottleneck" of metagenomic efforts. While this approach is more reliable for CAZyme screening, the available methods are time-consuming and should be assayed using a massive and an expensive number of polymeric substrates.

On the other hand, the sequence-based metagenomic approach seems to be more feasible, but the discovery of complete genes from environmental samples has been limited by the difficult-to-reach full-length genes and the classification of these genes to CAZy families. Additionally, the majority of CAZy families is frequently polyspecific, containing enzymes acting on different substrates or generating different products. For example, 20 EC numbers are classified into the family GH5, such as chitosanase (EC 3.2.1.132), lichenase (EC 3.2.1.73), beta-glucosidase (3.2.1.21), and endo-beta-1,4-xylanase (3.2.1.-). This complexity of CAZy families requires functional studies to carefully address a new gene/protein into a specific family. A sequence-based approach allows only the identification of CAZy-conserved domains, but it is not possible to assign the precise function based only on databases.

New enzymes isolated from metagenomes are being applied to improve the enzymatic hydrolysis process. Alvarez and coworkers reported the characterization of a novel endoxylanase family GH10 isolated from the sugarcane soil metagenome (SCXyl). This GH10 was applied as a pretreatment step of phosphoric acid-pretreated sugarcane bagasse, before the addition of commercial cellulolytic cocktail, and significantly enhanced the saccharification process (Alvarez et al. 2013a). The GH11 xylanase from a compost-soil metagenome (rMxyl) (Verma et al. 2013) is another example of an enzyme isolated from metagenomes and applied to biomass conversion. The rMxyl is an enzyme with alkali- and thermostability that efficiently hydrolyzes wheat bran, corncobs, and sugarcane bagasse. Alkaline xylanases display better activity on agro-residues by decreasing the steric hindrance due to cellulose and increasing the level of solubility of hemicellulosic (Gruppen et al. 1992).

As mentioned above, a number of metagenome-derived CAZymes have been isolated from environmental samples and characterized. These "new" enzymes

Table 7.2 Contribution of metagenomic approaches to the structural biology of CAZymes

Protein name	CAZy family	EC number	Organism	PDB/3D
Chain A, Multicopper Oxidase Mglac	AA1	–	Uncultured bacteria	4E9W; 4E9X; 4E9Y
Laccase (Lac15)	AAnc	1.10.3.2	Uncultured bacterium	4F7K
Carbohydrate binding protein (Pel10)	CBM35	–	Uncultured bacterium	2W3J
vCBM60 (Cbm60)	CBM60	–	Uncultured bacterium	2XFD; 2XFE; 2XHH; 2XHJ
β-Glucosidase (Bgl1)	GH1	3.2.1.21	Uncultured bacterium	3CMJ, 3FIY, 3FIZ, 3FJ0, 4HZ6, 4HZ7, 4HZ8
β-Glucosidase (Td2F2)	GH2	3.2.1.21	Unidentified	3WH5, 3WH7, 3WH8, 3WH6
β-Glucosidase (JMB19063)	GH3	3.2.1.21	Compost metagenome	3U48, 3U4A, 3U48
Endoglucanase (RBcel1)	GH5	3.2.1.4	Uncultured bacterium	4EE9, 4M24
Endoglucanase/cellulase (cMGL504) (Cel5A)	GH5	3.2.1.4	Uncultured bacterium	4HTY, 4HU0
Cellulase (CelE1) (partial)	GH5	3.2.1.4	Uncultured bacterium	4M1R[A,B]
Cellulase (Lc-CelG)	GH9	3.2.1.4	Uncultured bacterium	3X17
Xylanase (SCXyl)	GH10	3.2.1.8	Uncultured bacterium	4K68
Xylanase A (Xyl-Orf19)	GH10	3.2.1.8	Uncultured bacterium	4HU8
Xylanase (EvXyn11fl) (Xyn11)	GH11	3.2.1.8	Uncultured bacterium	2VUJ, 2VUL
Cellulase (Lc-CelA)	GH12	3.2.1.4	Uncultured bacterium	3WX5
β-1,3-1,4-Glucanase/lichenase (Bgl14A1)	GH16	–	Uncultured murine large bowel bacterium	3I4I
Exo-α-1,5-L-arabinanase (Arn3;ARN3)	GH43	3.2.1.-	Uncultured bacterium	4KCB
Endo-α-1,5-L-arabinanase (Arn2;ARN2)	GH43	3.2.1.99	Uncultured rumen bacterium	4KCA
β-Xylosidase (RS223-BX)	GH43	3.2.1.37	Uncultured organism	4MLG
Bifunctional glucanase-xylanase (CelM2)	GH44	3.2.1.-	Uncultured bacterium	3FW6, 3II1

The data were based on the CAZy database (www.cazy.org)

exhibit special characteristics such as a high tolerance to acidic or alkaline solutions and resistance to high temperatures. However, metagenomic approaches have also been important for the field of structural biology. The number of structures of metagenome-derived CAZymes is far lower than the number of enzymes characterized at a biochemical level. By mining the entire CAZy database, very few structures with PDB entries (Protein Data Bank), are available, such as 2 AAs, 2 CBMs, and 16 GHs, which are described on (Table 7.2). There are no PDB entries for PLs, GTs, and CEs derived from metagenomes.

The possibility of exploring the genetic information of uncultured microorganisms is very promising from a structural point of view and surely enables the discovery of new domains, new structural motifs, and, therefore, new families. In 2013, Han reported a crystal structure of a termite gut bacterial xylanase. The protein contains a GH10 catalytic domain and an accessory domain, named Big_2 (immunoglobulin-like). The kinetic parameters, pH, and temperature profiles of the recombinant enzyme without the noncatalytic domain were different from the full-length enzyme and the former showed low catalytic activity, which suggests the noncatalytic domain could affect the biochemical and biophysical properties of the enzyme as well as the role of localization (Han et al. 2013).

7.4 Application of Metagenomics for Improvement of the Sustainability of an Ethanol Production Plant

As previously mentioned, the new Brazilian model of energy production from biomass (1G2G ethanol) represents one of the most sustainable and consistent models to meet the world's growing energy demand. However, in parallel to the increased ethanol production, the generation of residual liquid streams from the production process will also increase at a large scale.

The sugarcane vinasse (1G vinasse) is a voluminous wastewater generated in the distillation process at biorefineries and has high concentrations of organic matter and macro- and micronutrients. From an economic perspective, the best option for vinasse reuse is for the fertilization of crops (sugarcane). In this use case, the vinasse content (nutrients and high water content, above 90%) is returned to the soil by irrigation. However, improper vinasse disposal on sugarcane crops, i.e., overdoses of vinasse application and the lack of technical criteria in the reuse systems, decreases not only the productivity of soil but also the quality of superficial and groundwater that surrounds the irrigated area. In addition to the sugarcane 1G vinasse, pentose liquor (5C liquor) and 2G vinasse will be generated at a large scale. The 5C liquor is a residual liquid stream enriched with monosaccharides that contain five carbons in their structure and derive from the lignocellulosic material (i.e., bagasse and straw) pretreatment.

Regarding the vinasse generated in the production of ethanol from sugarcane bagasse (2G vinasse), it is known that its nutrient and mineral content is fairly low (Groposo Silveira et al. 2012); thus, unlike 1G sugarcane vinasse, its application in the soil as a biofertilizer is not justified. The situation is even worse when considering that 5C liquor would be present (Boussarsar et al. 2009; Rabelo 2010), which is composed of hemicellulose with xylose being the main reducing sugar. Regardless, whatever the type of reuse utilized for those residual liquid streams, the sustainability of ethanol plants is recognized to be important.

In this context, anaerobic digestion has become a requirement for the integration of the (1) power and steam cogeneration, (2) water use (collection, treatment, and disposal), and (3) waste disposal systems. Therefore, anaerobic digestion is a sustainable option for alleviating environmental concerns of such streams to energy recovery as biogas (hydrogen and methane).

Few studies have reported the potential to produce hydrogen (Ferraz Júnior et al. 2014; Santos et al. 2014; Lazaro et al. 2014; Ferraz Júnior et al. 2015a) and methane (Souza et al. 1992; Mota et al. 2013; Ferraz Júnior et al. 2016) using sugarcane vinasse 1G as a substrate. However, the gaps in information regarding the anaerobic digestion of vinasse 2G and 5C liquor hinder the implementation and expansion of this technology. Furthermore, information is also lacking concerning the biogas-producing microorganisms (BPM) that are extremely important in supporting technologies that aim to increase the sustainability of ethanol plants.

In this section, a case study is presented to foster a clearer understanding of the organization and behavior of heterogeneous communities using NGS (454-pyrosequencing) using a case study of two lab-scale up-flow anaerobic packed-bed reactors (APBR) applied to produce hydrogen using raw sugarcane vinasse as a substrate under mesophilic and thermophilic conditions. We also speculate as to how this information could help to optimize the process and reach a stable operation.

7.4.1 Case Study

Anaerobic digestion is a well-known technology that constitutes an alternative to sugarcane vinasse. It is based on the principles of hydrolysis, acidogenesis, acetogenesis, and methanogenesis. Furthermore, anaerobic digestion not only alleviates environmental concerns but also converts sugarcane vinasse into two categories of valuable products: biogas, a renewable fuel, and digested vinasse, which has improved fertilizer characteristics.

In addition to conventional anaerobic digestion from a combustible biogas (CH_4), the production of hydrogen from 1G vinasse may become very attractive for the following reasons: (1) hydrogen is a carbon-neutral energy carrier that is 2.75 times more energetic than fossil fuels due to its high energy yield (142.9 kJ g^{-1}); (2) sugarcane vinasse is a voluminous wastewater with high organic matter content; (3) anaerobic digestion technology requires simple configurations of reactors and less input energy compared with other technologies; (4) the process allows for the sustainability of ethanol plants via water use and disposal systems.

To exemplify the use of metagenomics tool in the field of anaerobic digestion of sugarcane vinasse, a case study considering previous studies (Ferraz Júnior et al. 2015a, b) was compiled and compared using principal coordinates analysis (PCoA) and PAST Software. Based on Ferraz Júnior (2015a, b), samples were distributed along the bed zone of mesophilic and thermophilic APBR, and the samples were collected from five sampling points. A compiled sample of all of the subsamples was used for the analysis. A 16S rRNA 454-pyrosequence analysis was performed on samples taken from the mesophilic and thermophilic APBR filled with low-density polyethylene (LDP) during a period of high hydrogen production (mesophilic and thermophilic, MI and TI, respectively) and during a period without hydrogen production (MII and TII, respectively).

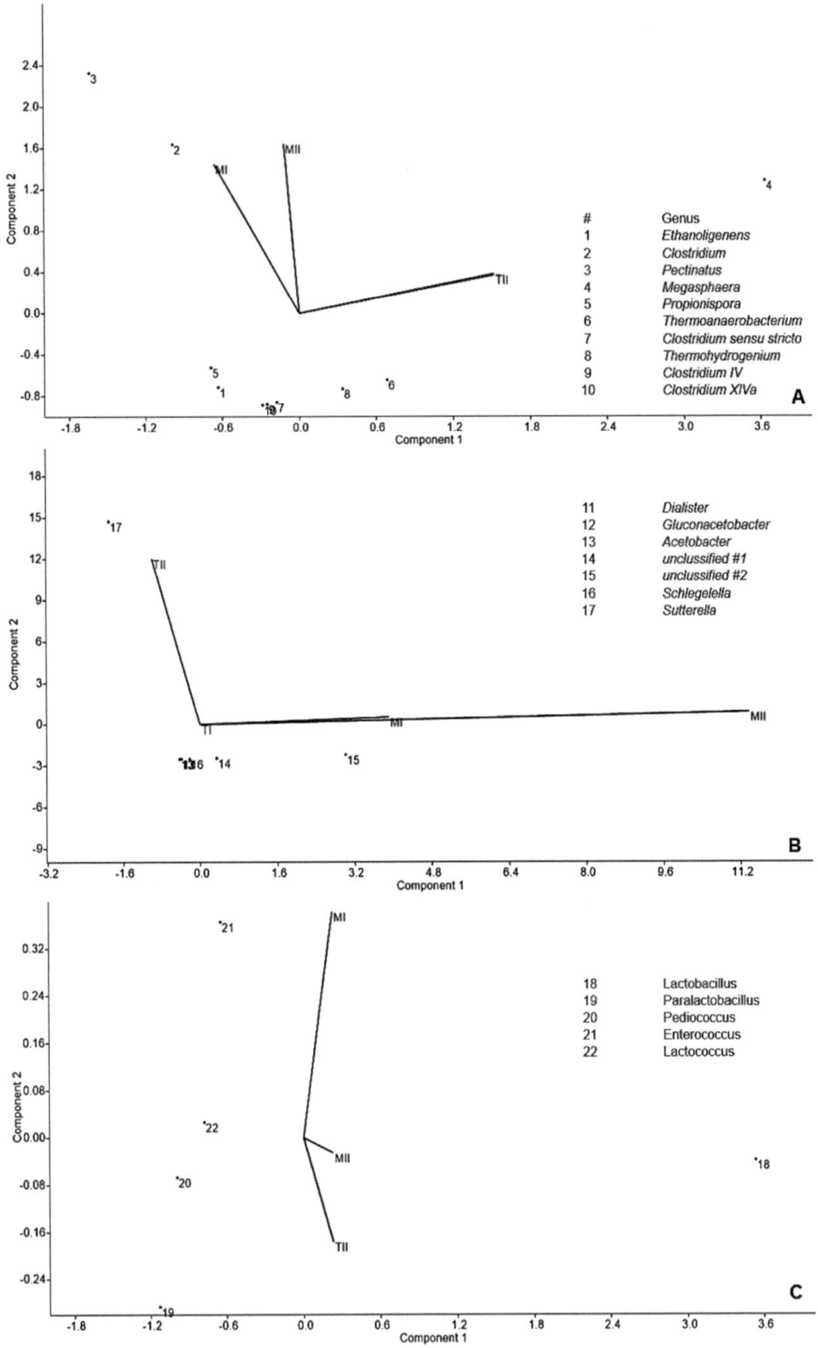

Fig. 7.2 Principal component analysis (PCA) of 16S rRNA of the microbial community in up-flow anaerobic packed-bed reactors (APBR) applied to hydrogen production using raw sugarcane vinasse as a substrate. (**a**) Hydrogen-producing bacteria (HPB); (**b**) non-hydrogen-producing bacteria (NHPB); (**c**) lactic acid-producing bacteria (LAPB), in which MI is mesophilic APBR at the tenth day of operation, MII is mesophilic APBR at the 30th day of operation, TI is thermophilic APBR at the 30th day of operation, and TII is thermophilic APBR at the 60th day of operation

In those studies, approximately 4000 reads were obtained for each sample. The reads obtained from samples MI and MII were classified into seven phylogenetic groups according to the RDP pipeline: *Coriobacteriales*, *Bifidobacteriales*, *Bacteroidales*, *Rhodospirillales*, *Lactobacillales*, *Clostridiales*, and *Burkholderiales*. The three former families were also detected in samples TI and TII in addition to *Thermoanaerobacterales* and *Selenomonadales*. The role of microorganisms in mesophilic and thermophilic APBR using raw sugarcane vinasse to produce hydrogen can be summarized in hydrogen-producing bacteria (HPB) (Fig. 7.2a), lactic acid-producing bacteria (LAPB) (Fig. 7.2b), and non-hydrogen-producing bacteria (NHPB) (Fig. 7.2c).

Thermophilic fermentation greatly favors HPB diversity compared with mesophilic fermentation. Furthermore, the relative abundance of HPB did not change substantially during the thermophilic operation, indicating their persistence in the reactor during the entire operation (TI and TII) (Fig. 7.2a). In Fig. 7.2b, sequences affiliated with the family *Comamonadaceae* (*unclassified* #1) and the order *Burkholderiales* (*unclassified* #2) were dominant after 30 days in the mesophilic APBR operation (MII). These microorganisms are known as NHPB and are considered to be efficient in organic matter degradation in the presence of oxygen (Niemi et al. 2009). This family has been detected in hydrogen-producing reactors, and their persistence has been associated with microaerophilic conditions. There is disagreement in the literature about the role of LAPB in hydrogen-producing systems (Fig. 7.2c). Some of them (i.e., LAB) can be highlighted as having (1) an inhibitory effect of LAPB on hydrogen production (Noike 2002; Jo et al. 2007); (2) production of small amounts of hydrogen by medium acidification (Li and Fang 2007); and (3) substrate competition between LAPB and hydrogen-producing bacteria leading to system instability (Ohnishi et al. 2010). However, the LAPB in the thermophilic APBR represented 46.7% of the reads at the 30th day of operation, indicating that the LAPB were associated with hydrogen production.

From the mesophilic APBR monitoring, the low hydrogen production and yield achieved were associated with (1) microaerophilic conditions (*Burkholderiales* as indicator microorganisms and oxireduction potential of 55.7 ± 23 mV) that inhibited the HPM growth and favored the NHPM growth and (2) excess nitrogen in the sugarcane vinasse composition that resulted in excessive biomass growth, diverting the hydrogen pathway to uptake and cell growth. The combination of these two factors severely affected the mesophilic APBR operation, leading to a decay in hydrogen production (Ferraz Júnior et al. 2015a).

These analyses drove changes in the APBR operation to improve the biological process. The adoption of a thermophilic operation, rather than a mesophilic operation, enabled the continuous and stable production of hydrogen from raw sugarcane vinasse. The increase in temperature decreased both the biomass yield and the oxygen solubility. According to Ferraz Júnior et al. (2014), applying a higher organic loading rate (OLR) of 84.2 kg-COD m^{-3} d^{-1} resulted in maximum volumetric hydrogen production (VHP) and a yield of 2283.8 mL-H_2 d^{-1} $L^{-1}_{reactor}$ and 3.7 mol-H_2 mol^{-1} total carbohydrates, respectively. Moreover, throughout the thermophilic APBR operation, NHPB (*Sutterella*, order *Burkholderiales*) increased in the reactor and reached 17.3% of the reads at the 60th day of operation (Fig. 7.2b). This value is three times lower than those computed in a mesophilic APBR at the 30th day of operation. Collectively, this finding reinforced the advantages of thermophilic fermentation, including the ability to decrease the inhibition of autochthonous microorganisms that are typically present in the inflow. Finally, the thermophilic operation represented a great and rational option for sugarcane vinasse that is discharged at high temperatures (85–90 °C) and, thus, does not require energy input.

7.5 Concluding Remarks

Metagenomic studies based on sugarcane and applied to the development of biofuels and environmental sustainability at biorefineries have been performed to increase access to the genetic information of microbial communities and uncultured microorganisms. In this chapter, we have described the contribution of metagenomics to the identification of novel genes coding for proteins involved in plant biomass degradation, as well as in the assessment of anaerobic process by specialized microbial communities for bioenergy production. We expect that this chapter will provide insight for future studies on metagenomics applied to biomass-to-bioenergy applications.

Acknowledgements The authors gratefully acknowledge financial support from FAPESP (Thematic Project 2009/15984-0 and Postdoctoral Project 2013/15665-8, 2012/20549-4 to A.R.L.D.).

References

Adsul MG, Singhvi MS, Gaikaiwari SA, Gokhale DV (2011) Development of biocatalysts for production of commodity chemicals from lignocellulosic biomass. Bioresour Technol 102:4304–4312. doi:10.1016/j.biortech.2011.01.002

Alvarez TM, Goldbeck R, dos Santos CR et al (2013a) Development and biotechnological application of a novel endoxylanase family GH10 identified from sugarcane soil metagenome. PLoS One 8:e70014. doi:10.1371/journal.pone.0070014

Alvarez TM, Paiva JH, Ruiz DM et al (2013b) Structure and function of a novel cellulase 5 from sugarcane soil metagenome. PLoS One 8:e83635. doi:10.1371/journal.pone.0083635

Amann RI, Ludwig W, Schleifer KH (1995) Phylogenetic identification and in situ detection of individual microbial cells without cultivation. Microbiol Rev 59:143–169

Barba M, Czosnek H, Hadidi A (2014) Historical perspective, development and applications of next-generation sequencing in plant virology. Viruses 6:106–136. doi:10.3390/v6010106

Bastien G, Arnal G, Bozonnet S et al (2013) Mining for hemicellulases in the fungus-growing termite Pseudacanthotermes militaris using functional metagenomics. Biotechnol Biofuels 6:78. doi:10.1186/1754-6834-6-78

Berghem LE, Pettersson LG, Axio-Fredriksson UB (1976) The mechanism of enzymatic cellulose degradation. Purification and some properties of two different 1,4beta-glucan glucanohydrolases from Trichoderma viride. Eur J Biochem 61:621–630

Berlemont R, Delsaute M, Pipers D et al (2009) Insights into bacterial cellulose biosynthesis by functional metagenomics on Antarctic soil samples. ISME J 3:1070–1081. doi:10.1038/ismej.2009.48

Bertrand H, Poly F, Van VT et al (2005) High molecular weight DNA recovery from soils prerequisite for biotechnological metagenomic library construction. J Microbiol Methods 62:1–11. doi:10.1016/j.mimet.2005.01.003

Boussarsar H, Rogé B, Mathlouthi M (2009) Optimization of sugarcane bagasse conversion by hydrothermal treatment for the recovery of xylose. Bioresour Technol 100:6537–6542. doi:10.1016/j.biortech.2009.07.019

Brennan Y, Callen WN, Christoffersen L et al (2004) Unusual microbial xylanases from insect guts. Appl Environ Microbiol 70:3609–3617. doi:10.1128/AEM.70.6.3609-3617.2004

Brenner K, You L, Arnold FH (2008) Engineering microbial consortia: a new frontier in synthetic biology. Trends Biotechnol 26:483–489. doi:10.1016/j.tibtech.2008.05.004

Buermans HPJ, den Dunnen JT (2014) Next generation sequencing technology: advances and applications. Biochim Biophys Acta 1842:1932–1941. doi:10.1016/j.bbadis.2014.06.015

Carpita NC, McCann MC (2000) The cell wall. In: Buchanan BB, Gruissem W, Jones R (eds) Biochemistry and molecular biology of plants. American Society of Plant Physiologists, Rockville, pp 52–109

Daniel R (2005) The metagenomics of soil. Nat Rev Microbiol 3:470–478. doi:10.1038/nrmicro1160

De Souza AP, Grandis A, Leite DCC, Buckeridge MS (2014) Sugarcane as a bioenergy source: history, performance, and perspectives for second-generation bioethanol. Bioenergy Res 7:24–35. doi:10.1007/s12155-013-9366-8

De Souza AP, Leite DCC, Pattathil S et al (2013) Composition and structure of sugarcane cell wall polysaccharides: implications for second-generation bioethanol production. Bioenergy Res 6:564–579. doi:10.1007/s12155-012-9268-1

Dröge J, Mchardy AC (2012) Taxonomic binning of metagenome samples generated by next-generation sequencing technologies. Brief Bioinform 13:646–655. doi:10.1093/bib/bbs031

Du Toit PJ, Olivier SP, van Biljon PL (1984) Sugar cane bagasse as a possible source of fermentable carbohydrates. I. Characterization of bagasse with regard to monosaccharide, hemicellulose, and amino acid composition. Biotechnol Bioeng 26:1071–1078. doi:10.1002/bit.260260909

Duan CJ, Xian L, Zhao GC et al (2009) Isolation and partial characterization of novel genes encoding acidic cellulases from metagenomes of buffalo rumens. J Appl Microbiol 107:245–256. doi:10.1111/j.1365-2672.2009.04202.x

El-Metwally S, Ouda MO, Helmy M (2014) Next generation sequencing technologies and challenges in sequence assembly. Springer-Verlag, New York

Elend C, Schmeisser C, Leggewie C et al (2006) Isolation and biochemical characterization of two novel metagenome-derived esterases. Appl Environ Microbiol 72:3637–3645. doi:10.1128/AEM.72.5.3637-3645.2006

Entcheva P, Liebl W, Johann A et al (2001) Direct cloning from enrichment cultures, a reliable strategy for isolation of complete operons and genes from microbial consortia. Appl Environ Microbiol 67:89–99. doi:10.1128/AEM.67.1.89-99.2001

Fang Z, Li T, Wang Q et al (2011) A bacterial laccase from marine microbial metagenome exhibiting chloride tolerance and dye decolorization ability. Appl Microbiol Biotechnol 89:1103–1110. doi:10.1007/s00253-010-2934-3

Feng Y, Duan CJ, Pang H et al (2007) Cloning and identification of novel cellulase genes from uncultured microorganisms in rabbit cecum and characterization of the expressed cellulases. Appl Microbiol Biotechnol 75:319–328. doi:10.1007/s00253-006-0820-9

Ferraz Júnior ADN, Etchebehere C, Zaiat M (2015a) High organic loading rate on thermophilic hydrogen production and metagenomic study at an anaerobic packed-bed reactor treating a residual liquid stream of a Brazilian biorefinery. Bioresour Technol 186:81–88. doi:10.1016/j.biortech.2015.03.035

Ferraz Júnior ADN, Etchebehere C, Zaiat M (2015b) Mesophilic hydrogen production in acidogenic packed-bed reactors (apbr) using raw sugarcane vinasse as substrate: influence of support materials. Anaerobe 34:94–105. doi:10.1016/j.anaerobe.2015.04.008

Ferraz Júnior ADN, Koyama MH, de Araújo Júnior MM, Zaiat M (2016) Thermophilic anaerobic digestion of raw sugarcane vinasse. Renew Energy 89:245–252. doi:10.1016/j.renene.2015.11.064

Ferraz Júnior ADN, Wenzel J, Etchebehere C, Zaiat M (2014) Effect of organic loading rate on hydrogen production from sugarcane vinasse in thermophilic acidogenic packed bed reactors. Int J Hydrogen Energy 39:16852–16862. doi:10.1016/j.ijhydene.2014.08.017

Gabor EM, De Vries EJ, Janssen DB (2003) Efficient recovery of environmental DNA for expression cloning by indirect extraction methods. FEMS Microbiol Ecol 44:153–163. doi:10.1016/S0168-6496(02)00462-2

Gabor EM, De Vries EJ, Janssen DB (2004) Construction, characterization, and use of small-insert gene banks of DNA isolated from soil and enrichment cultures for the recovery of novel amidases. Environ Microbiol 6:948–958. doi:10.1111/j.1462-2920.2004.00643.x

Goldemberg J, Coelho ST, Nastari PM, Lucon O (2004) Ethanol learning curve—the Brazilian experience. Biomass Bioenergy 26:301–304. doi:10.1016/S0961-9534(03)00125-9

Grada A, Weinbrecht K (2013) Next-generation sequencing: methodology and application. J Invest Dermatol 133:e11–e11. doi:10.1038/jid.2013.248

GranBio (2015) GranBio. http://www.granbio.com.br/. Accessed 21 Feb 2015

Grant S, Sorokin DY, Grant WD et al (2004) A phylogenetic analysis of Wadi el Natrun soda lake cellulase enrichment cultures and identification of cellulase genes from these cultures. Extremophiles 8:421–429. doi:10.1007/s00792-004-0402-7

Groposo Silveira CJ, Melo Santa Anna LM, Machado de Castro A, et al (2012) Method for producing energy-rich gases from lignocellulosic material streams

Gruppen H, Hamer RJ, Voragen AGJ (1992) Water unextractable cell wall material from wheat flour. 2. Fractionation of alkali extracted polymers and comparison with water extractable arabinoxylans. J Cereal Sci 16:53–67

Han Q, Liu N, Robinson H et al (2013) Biochemical characterization and crystal structure of a GH10 xylanase from termite gut bacteria reveal a novel structural feature and significance of its bacterial Ig-like domain. Biotechnol Bioeng 110:3093–3103. doi:10.1002/bit.24982

Handelsman J (2004) Metagenomics: application of genomics to uncultured microorganisms. Microbiol Mol Biol Rev 68:669–685. doi:10.1128/MMBR.68.4.669-685.2004

Handelsman J, Rondon MR, Brady SF et al (1998) Molecular biological access to the chemistry of unknown soil microbes: a new frontier for natural products. Chem Biol 5:R245–R249. doi:10.1016/S1074-5521(98)90108-9

Henrissat B (1991) A classification of glycosyl hydrolases based on amino acid sequence similarities. Biochem J 280(Pt 2):309–316

Hess M, Sczyrba A, Egan R et al (2011) Metagenomic discovery of biomass-degrading genes and genomes from cow rumen. Science 331:463–467. doi:10.1126/science.1200387

Himmel ME, Ding S-Y, Johnson DK et al (2007) Biomass recalcitrance: engineering plants and enzymes for biofuels production. Science 315:804–807. doi:10.1126/science.1137016

Holben WE, Jansson JK, Chelm BK, Tiedje JM (1988) DNA probe method for the detection of specific microorganisms in the soil bacterial community. Appl Environ Microbiol 54:703–711

Hu Y, Zhang G, Li A et al (2008) Cloning and enzymatic characterization of a xylanase gene from a soil-derived metagenomic library with an efficient approach. Appl Microbiol Biotechnol 80:823–830. doi:10.1007/s00253-008-1636-6

Jeong YS, Na HB, Kim SK et al (2012) Characterization of xyn10J, a novel family 10 xylanase from a compost metagenomic library. Appl Biochem Biotechnol 166:1328–1339. doi:10.1007/s12010-011-9520-8

Jo JH, Jeon CO, Lee DS, Park JM (2007) Process stability and microbial community structure in anaerobic hydrogen-producing microflora from food waste containing kimchi. J Biotechnol 131:300–308. doi:10.1016/j.jbiotec.2007.07.492

Khademi S, Zhang D, Swanson SM et al (2002) Determination of the structure of an endoglucanase from Aspergillus niger and its mode of inhibition by palladium chloride. Acta Crystallogr D Biol Crystallogr 58:660–667

Kim S-Y, Oh D-B, Kwon O (2014) Characterization of a lichenase isolated from soil metagenome. J Microbiol Biotechnol 24:1699–1706. doi:10.4014/jmb.1406.06012

Knietsch A, Waschkowitz T, Bowien S et al (2003) Construction and screening of metagenomic libraries derived from enrichment cultures: generation of a gene bank for genes conferring alcohol oxidoreductase activity on Escherichia coli. Appl Environ Microbiol 69:1408–1416. doi:10.1128/AEM.69.3.1408-1416.2003

Lazaro CZ, Perna V, Etchebehere C, Varesche MBA (2014) Sugarcane vinasse as substrate for fermentative hydrogen production: the effects of temperature and substrate concentration. Int J Hydrogen Energy 39:6407–6418. doi:10.1016/j.ijhydene.2014.02.058

Lee CC, Kibblewhite-Accinelli RE, Wagschal K et al (2006) Cloning and characterization of a cold-active xylanase enzyme from an environmental DNA library. Extremophiles 10:295–300. doi:10.1007/s00792-005-0499-3

Lee CM, Lee YS, Seo SH et al (2014) Screening and characterization of a novel cellulase gene from the gut microflora of Hermetia illucens using metagenomic library. J Microbiol Biotechnol 24:1196–1206. doi:10.4014/jmb.1405.05001

Leff LG, Dana JR, McArthur JV, Shimkets LJ (1995) Comparison of methods of DNA extraction from stream sediments. Appl Environ Microbiol 61:1141–1143 7793915

Levasseur A, Drula E, Lombard V et al (2013) Expansion of the enzymatic repertoire of the CAZy database to integrate auxiliary redox enzymes. Biotechnol Biofuels 6:41. doi:10.1186/1754-6834-6-41

Levasseur A, Piumi F, Coutinho PM et al (2008) FOLy: an integrated database for the classification and functional annotation of fungal oxidoreductases potentially involved in the degradation of lignin and related aromatic compounds. Fungal Genet Biol 45:638–645

Li C, Fang HHP (2007) Fermentative hydrogen production from wastewater and solid wastes by mixed cultures. Crit Rev Environ Sci Technol 37:1–39

Li L-L, McCorkle SR, Monchy S et al (2009) Bioprospecting metagenomes: glycosyl hydrolases for converting biomass. Biotechnol Biofuels 2:10. doi:10.1186/1754-6834-2-10

Ling LL, Schneider T, Peoples AJ et al (2015) A new antibiotic kills pathogens without detectable resistance. Nature 517:455–459. doi:10.1038/nature14098

Lombard V, Golaconda Ramulu H, Drula E et al (2014) The carbohydrate-active enzymes database (CAZy) in 2013. Nucleic Acids Res 42:D490–D495. doi:10.1093/nar/gkt1178

Mackelprang R, Waldrop MP, DeAngelis KM et al (2011) Metagenomic analysis of a permafrost microbial community reveals a rapid response to thaw. Nature 480:368–371. doi:10.1038/nature10576

MacLean D, Jones JDG, Studholme DJ (2009) Application of "next-generation" sequencing technologies to microbial genetics. Nat Rev Microbiol 7:287–296. doi:10.1038/nrmicro2088

Mardis ER (2008) The impact of next-generation sequencing technology on genetics. Trends Genet 24:133–141. doi:10.1016/j.tig.2007.12.007

Martin M, Biver S, Steels S et al (2014) Identification and characterization of a halotolerant, cold-active marine endo-β-1,4-glucanase by using functional metagenomics of seaweed-associated microbiota. Appl Environ Microbiol 80:4958–4967. doi:10.1128/AEM.01194-14

Mhuantong W, Charoensawan V, Kanokratana P et al (2015) Comparative analysis of sugarcane bagasse metagenome reveals unique and conserved biomass-degrading enzymes among lignocellulolytic microbial communities. Biotechnol Biofuels 8:16. doi:10.1186/s13068-015-0200-8

Milanez A, Nyko D, Garcia J, Dos Reis BS (2012) O déficit de produção de etanol no Brasil entre 2012 e 2015: determinantes, consequências e sugestões de política. BNDES Setorial 35:277–302

Mota VT, Santos FS, Amaral MCS (2013) Two-stage anaerobic membrane bioreactor for the treatment of sugarcane vinasse: assessment on biological activity and filtration performance. Bioresour Technol 146:494–503. doi:10.1016/j.biortech.2013.07.110

Nass LL, Pereira PAA, Ellis D (2007) Biofuels in Brazil: an overview. Crop Sci 47:2228–2237. doi:10.2135/cropsci2007.03.0166

Nichols D, Cahoon N, Trakhtenberg EM et al (2010) Use of ichip for high-throughput in situ cultivation of "uncultivable" microbial species. Appl Environ Microbiol 76:2445–2450. doi:10.1128/AEM.01754-09

Niemi RM, Heiskanen I, Heine R, Rapala J (2009) Previously uncultured beta-Proteobacteria dominate in biologically active granular activated carbon (BAC) filters. Water Res 43:5075–5086. doi:10.1016/j.watres.2009.08.037

Noike T (2002) Inhibition of hydrogen fermentation of organic wastes by lactic acid bacteria. Int J Hydrogen Energy 27:1367–1371. doi:10.1016/S0360-3199(02)00120-9
Ogram A, Sayler GS, Barkay T (1987) The extraction and purification of microbial DNA from sediments. J Microbiol Methods 7:57–66. doi:10.1016/0167-7012(87)90025-X
Ohnishi A, Bando Y, Fujimoto N, Suzuki M (2010) Development of a simple bio-hydrogen production system through dark fermentation by using unique microflora. Int J Hydrogen Energy 35:8544–8553. doi:10.1016/j.ijhydene.2010.05.113
Pace NR, Stahl DA, Lane DJ, Olsen GJ (1986) The analysis of natural microbial populations by ribosomal RNA sequences. Adv Microb Ecol 9:1–55
Pandhal J, Noirel J (2014) Synthetic microbial ecosystems for biotechnology. Biotechnol Lett 36:1141–1151. doi:10.1007/s10529-014-1480-y
Rabelo SC (2010) Avaliação E Otimização De Pré-Tratamentos E Hidrólise Enzimática Do Bagaço De Cana-De-Açúcar Para a Produção De Etanol De Segunda Geração. p 447
Raízen (2015) Raízen. http://www.raizen.com.br/. Accessed 10 Jan 2015
Rees HC, Grant S, Jones B et al (2003) Detecting cellulase and esterase enzyme activities encoded by novel genes present in environmental DNA libraries. Extremophiles 7:415–421. doi:10.1007/s00792-003-0339-2
Saiki RK, Gelfand DH, Stoffel S, Scharf SJ, Higuchi R, Horn GT, Mullis KB, HA E (1988) Primer-directed enzymatic amplification of DNA with a thermostable DNA polymerase. Science 239:487–491
Sanger F, Nicklen S, Coulson AR (1977) DNA sequencing with chain-terminating inhibitors. Proc Natl Acad Sci U S A 74:5463–5467. doi:10.1073/pnas.74.12.5463
Santos SC, Ferreira Rosa PR, Sakamoto IK et al (2014) Continuous thermophilic hydrogen production and microbial community analysis from anaerobic digestion of diluted sugar cane stillage. Int J Hydrogen Energy 39:9000–9011. doi:10.1016/j.ijhydene.2014.03.241
Schloss PD, Handelsman J (2003) Biotechnological prospects from metagenomics. Curr Opin Biotechnol 14:303–310. doi:10.1016/S0958-1669(03)00067-3
Segato F, Damásio AR, Squina FM, Prade RA (2014) Genomics review of holocellulose deconstruction by aspergilli. Microbiol Mol Bio Rev 78:588–613
Simon C, Daniel R (2011) Metagenomic analyses: past and future trends. Appl Environ Microbiol 77:1153–1161. doi:10.1128/AEM.02345-10
Singh R, Dhawan S, Singh K, Kaur J (2012) Cloning, expression and characterization of a metagenome derived thermoactive/thermostable pectinase. Mol Biol Rep 39:8353–8361. doi:10.1007/s11033-012-1685-x
Souza ME, Fuzaro G, Polegato AR (1992) Thermophilic anaerobic digestion of vinasse in pilot plant UASB reactor. Water Sci Technol 25:213–222
Staley A, Konopka JT (1985) Measurement of in situ activities of nonphotosynthetic microorganisms in aquatic and terrestrial habitats. Annu Rev Microbiol 39:321–346
Steffan RJ, Goksøyr J, AK B, RM A (1988) Recovery of DNA from soils and sediments. Appl Environ Microbiol 54:2908–2915
Stroobants A, Portetelle D, Vandenbol M (2014) New carbohydrate-active enzymes identified by screening two metagenomic libraries derived from the soil of a winter wheat field. J Appl Microbiol 117:1045–1055. doi:10.1111/jam.12597
Sweeney MD, Xu F (2012) Biomass converting enzymes as industrial biocatalysts for fuels and chemicals: recent developments. Catalysts 2:244–263. doi:10.3390/catal2020244
Tringe SG, Rubin EM (2005) Metagenomics: DNA sequencing of environmental samples. Nat Rev Genet 6:805–814. doi:10.1038/nrg1709
Tringe SG, von Mering C, Kobayashi A et al (2005) Comparative metagenomics of microbial communities. Science 308:554–557. doi:10.1126/science.1107851
Tyson GW, Chapman J, Hugenholtz P et al (2004) Community structure and metabolism through reconstruction of microbial genomes from the environment. Nature 428:37–43. doi:10.1038/nature02340
van der Lelie D, Taghavi S, McCorkle SM et al (2012) The metagenome of an anaerobic microbial community decomposing poplar wood chips. PLoS One 7:e36740. doi:10.1371/journal.pone.0036740

Venter JC, Remington K, Heidelberg JF et al (2004) Environmental genome shotgun sequencing of the Sargasso Sea. Science 304:66–74. doi:10.1126/science.1093857

Verma D, Kawarabayasi Y, Miyazaki K, Satyanarayana T (2013) Cloning, expression and characteristics of a novel alkalistable and thermostable xylanase encoding gene (Mxyl) retrieved from compost-soil metagenome. PLoS One 8:e52459. doi:10.1371/journal.pone.0052459

Voget S, Steele HL, Streit WR (2006) Characterization of a metagenome-derived halotolerant cellulase. J Biotechnol 126:26–36. doi:10.1016/j.jbiotec.2006.02.011

Xiang L, Li A, Tian C et al (2014) Identification and characterization of a new acid-stable endoglucanase from a metagenomic library. Protein Expr Purif 102:20–26. doi:10.1016/j.pep.2014.07.009

Yun J, Kang S, Park S et al (2004) Characterization of a novel amylolytic enzyme encoded by a gene from a soil-derived metagenomic library. Appl Environ Microbiol 70:7229–7235. doi:10.1128/AEM.70.12.7229-7235.2004

Part III
Microorganisms and Sugarcane Genetics

Chapter 8
Yeast for Pentose Fermentation: Isolation, Screening, Performance, Manipulation, and Prospects

Sandra Regina Ceccato-Antonini, Carolina Brito Codato, Cristina Martini, Reinaldo Gaspar Bastos, and Sâmia Maria Tauk-Tornisielo

Abstract The global demand for energy has led the research and the investments to use plant biomass to convert the sugars contained in this material into ethanol. The characteristics of the substrate and process have a strong impact on the choice of microorganisms to be used for fermentation of the sugars. In the most of feedstocks for ethanol production, the sugars containing five carbons (pentoses) are abundant. Naturally occurring yeasts that can use pentoses as carbon source have been isolated from the environment, and among them, *Pichia stipitis* is one of the most important species. However, some important characteristics needed in ethanol industry are high resistance to inhibiting compounds and high fermentation performance and, until this moment, none a single strain that gather these features has not been found naturally. Techniques of evolutionary engineering and genetic manipulation have been applied to introduce and select the required traits for pentose fermentation in *Saccharomyces cerevisiae*, the most employed yeast industrially. This chapter discusses the context of the microorganisms, especially the yeast group, in the fermentation of hemicellulosic substrates for bioethanol production regarding isolation, screening, performance, limitations, prospects, and state of the art, trying to contribute to the improvement of the global process of ethanol production.

Keywords Pentoses • Ethanol • Hemicelluloses • Yeasts • Fermentation

---

S.R. Ceccato-Antonini (✉) • C.B. Codato • R.G. Bastos
Department of Tecnologia Agroindustrial e Socio-Economia Rural, Centro de Ciencias Agrarias, Universidade Federal de São Carlos, Via Anhanguera, Km 174, 13600-970 Araras, SP, Brazil
e-mail: antonini@cca.ufscar.br

C. Martini • S.M. Tauk-Tornisielo
Centro de Estudos Ambientais, Universidade Estadual Paulista Julio de Mesquita Filho, Avenida 24-A, 1515, 13506-900 Rio Claro, SP, Brazil

M.S. Buckeridge, A.P. De Souza (eds.), *Advances of Basic Science for Second Generation Bioethanol from Sugarcane*, DOI 10.1007/978-3-319-49826-3_8

8.1 Introduction

Plant biomass has a great potential as a source of bioenergy. The main components of plant biomass are hemicellulose, cellulose, and lignin, highly resistant to degradation due to the complex chemical and physical structures (Buckeridge et al. 2015). Cellulose is a homopolymer of glucose, whereas hemicelluloses are heteropolymers of xylose, arabinose, mannose, galactose, glucose, and uronic acids (Hayn et al. 1993; Meshitsuka and Isogai 1996). (For details about cell wall structure and physical properties, see Chaps. 2 and 3.) To make the sugars from plant cell wall accessible to be converted into ethanol by microorganisms is not a simple task (De Souza et al. 2014). Steps before the fermentation involve pretreatment and hydrolysis with chemical and/or physical methods to separate the lignin from hemicelluloses and cellulose (Hendriks and Zeeman 2009; Margeot et al. 2009; Soccol et al. 2010; Laluce et al. 2012; Buckeridge et al. 2015).

In the commercial production of ethanol directly from sugars in the so-called first generation bioethanol—1G (Amorim et al. 2011), the most common microorganism used during fermentation is *Saccharomyces cerevisiae*. This yeast species, which is conventionally used in industrial processes, is unable to metabolize pentoses, such as xylose and arabinose. These sugars are abundant in hemicelluloses, which are a great proportion of the biomass (De Souza et al. 2013, 2015—see Chap. 2). Natural strains of pentose-fermenting microorganisms, especially yeasts, have been sought-after in many locations. A useful technique for finding yeast strains with specific properties has been the prospection in natural or industrial habitats (Ferreira et al. 2011). Efforts have also been made to construct recombinant yeast strains to enhance xylose and arabinose fermentation over the past few decades (Zhang and Geng 2012; Harner et al. 2015).

The conditions in which the production of ethanol from biomass, named second generation ethanol (2G), is carried out differ substantially from the ones utilized in the fermentation of sugarcane juice or molasses. The demand for oxygen, low tolerance to ethanol, and sensitivity to the inhibitors produced during the hydrolysis step are among the features of a given candidate microorganism for the fermentation process. As a result, reaching high yields of ethanol from pentoses on a large scale is still a challenge (Hahn-Hagerdal and Pamment 2004).

Is it possible to find a natural strain of microorganism able to efficiently ferment pentoses? Where should they be searched for? What are the conditions to be optimized in order to obtain higher ethanol yields using hemicellulose-rich biomass hydrolysates? Would it be more efficient to engineer microbial strains so that they could use pentoses efficiently? These questions rise insistently whenever this subject comes to discussion. Surely, the fermentation step is not the only Achilles heel in the context of 2G bioethanol production. The hydrolysis step displays some limitations both concerning the aggressiveness of the acid/alkali treatment resulting in loss of soluble fractions of the cell wall polysaccharides (De Souza et al. 2013) and high cost of enzyme utilization (Laluce et al. 2012).

This chapter discusses the role of the microorganisms in the fermentation of hemicellulosic substrates for bioethanol production, focusing on the yeast group. Here, the isolation, screening, performance, limitations, prospects, and state of the art are revised in an attempt to contribute to the improvement of the global process of 2G bioethanol production.

8.2 Exploiting the Natural Biodiversity to Screen for Pentose-Fermenting Yeasts

A wide diversity of microbial species, including the yeast group, plays crucial roles in biogeochemical cycles and in food, biotechnology, and pharmaceutical industries (Boundy-Mills 2006). The exploration of the diverse ecological niches of yeasts revealed the great diversity of species living in various habitats. Among the numerous species to be possibly discovered in specific and unusual habitats, many will be likely found to possess enzymes, metabolic routes, and display physiological properties valuable for biotechnological applications (Deak 2009).

In yeasts, the initial metabolic pathways for D-xylose, L-arabinose, and glucose utilization are illustrated in Fig. 8.1. A series of reduction and oxidation steps involving the cofactors $NAD(P)^+$/NAD(P)H drive the conversion of xylose and arabinose to D-xylulose 5-phosphate to connect it to the central metabolic pathways (Hahn-Hagerdal et al. 2007). Xylitol is the common compound to both xylose and arabinose catabolic pathways (Fonseca et al. 2007a). Xylitol and arabitol, five-carbon sugar alcohols used in the food industry as sweeteners and in the production of human therapeutics (Sasaki et al. 2012; Kordowska-Wiater 2015), can be produced from the pentoses as well as ethanol, one of the most important renewable fuel.

The first report of xylose fermentation by yeasts was documented by Karczewska (1959). In 1976, a major review pointed out that nearly 50% of the known yeast species were able to assimilate xylose, but not to ferment it (Barnett 1976). In 1981, Schneider et al. identified *Pachysolen tannophilus* as a yeast capable of fermenting xylose to ethanol under aerobic condition. Two years later, *Candida shehatae* was first described by du Preez and van der Walt (1983) as an ethanol-producing yeast via pentose fermentation, inclusively with a better yield than *P. tannophilus.*

By 1984, Toivola et al. carried out a screening program for xylose-fermenting yeasts using strains of 193 species of yeasts, including 173 yeast species contained in Barnett's taxonomic book of 1979s edition, both glucose-fermenting and xylose-assimilating, in addition to 20 species not listed in this taxonomy book. Only 25 yeast species were found to be more efficient xylose fermenters, and of these, 14 were representative of *Candida.* Strains of *Brettanomyces naardenensis*, *C. shehatae*, *Candida tenuis*, *P. tannophilus*, *Pichia segobiensis*, and *Pichia stipitis* produced more than 1 g of ethanol per liter in medium with 2% xylose. The authors highlighted the striking characteristic of these xylose-fermenting yeasts. They were all

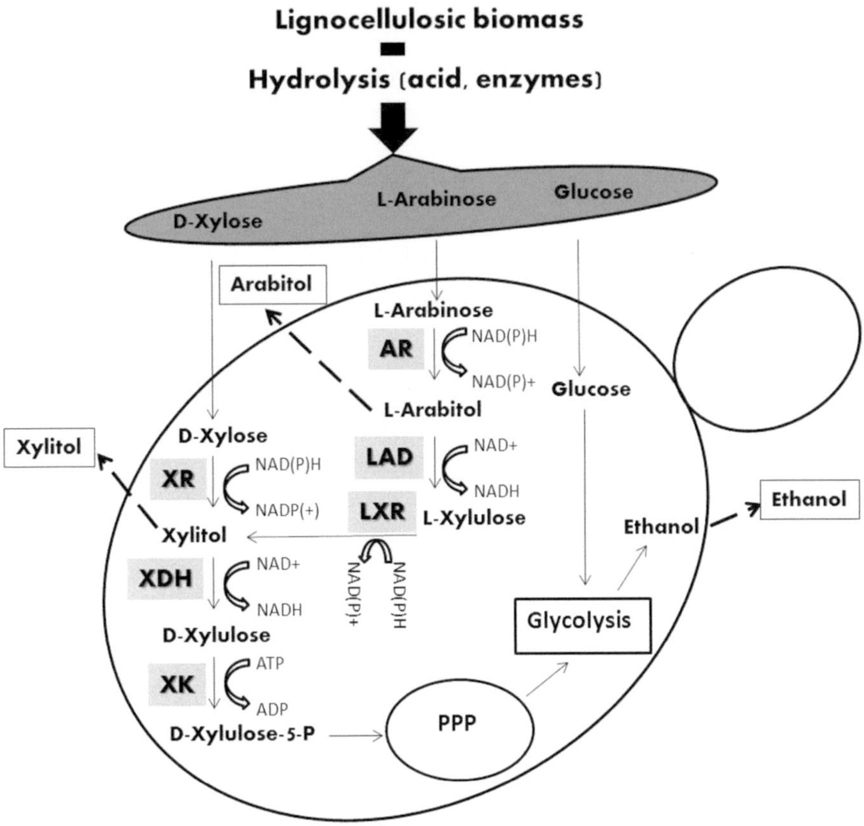

Fig. 8.1 Metabolic pathways of xylose, arabinose, and glucose, the main sugar products of hemicellulose hydrolysis of plant biomass, in yeast. *XR* xylose reductase, *XDH* xylitol dehydrogenase, *XK* xylulokinase, *AR* aldose reductase, *LAD* L-arabitol dehydrogenase, *LXR* L-xylulose reductase, *PPP* pentose phosphate pathway. Based on Hahn-Hagerdal et al. (2007), Laluce et al. (2012) and Wei et al. (2013)

isolated from wood-inhabiting insects, decaying wood, or other wood-related sources. Thus, they concluded that the xylose-fermenting yeasts have an ecological niche in such habitats.

Nigam et al. (1985) considered that the screening of xylose-fermenting yeasts should not be limited to culture collections and reported a study of the presence of xylose-fermenting yeasts in natural substrates likely to contain xylose. Among 412 isolates that were examined, 36 produced more than 1 g/L ethanol from 20 g/L xylose in oxygen non-limiting conditions. Also, six xylose-fermenting strains of a novel species of *Candida*, *C. lyxosophila*, were isolated from samples of woodland soil in Southern Africa (van der Walt et al. 1987).

In 1994, a review identified five yeast species of interest for xylose fermentation: *C. shehatae*, *C. tenuis*, *P. tannophilus*, *P. segobiensis*, and *P. stipitis* and mentioned

P. tannophilus and *P. stipitis* as the most extensively studied at that moment (Jeffries and Kurtzman 1994). *P. stipitis* is a naturally occurring xylose-fermenting yeast able to metabolize glucose, mannose, galactose, rhamnose, xylose, arabinose, cellobiose, and some lignin-related compounds (Jeffries and van Vleet 2009). It is still the most important natural yeast utilized for pentose fermentation and is also a source of genes to be introduced into *S. cerevisiae* strains by genetic manipulation.

Most of the yeast species found in soil can utilize aerobically arabinose, xylose, and cellobiose, which are the main products of the enzymatic hydrolysis of lignocellulosic plant material by bacteria and fungi, and assimilate intermediates of lignin degradation, all of them also found as components of root exudates (Tomme et al. 1995; Sampaio 1999; Fan et al. 2001; Botha 2006).

Pichia stipitis-like yeasts have been frequently isolated from wood-ingesting beetles over a wide area in the United States and Panama. A well-supported clade consisting of the passalid yeasts and *P. stipitis*, *P. segobiensis*, *C. shehatae*, and *Candida ergatensis* was distinguished, whose members were able to ferment and assimilate xylose or hydrolyse xylan, the major components of hemicelluloses of grasses (Carpita and Gibeaut 1993; Suh et al. 2003).

In the study by Rao et al. (2008), from 374 yeast strains isolated from rotten fruits and barks of trees, only 27 produced ethanol in the range of 0.12–0.38 g ethanol/g xylose.

Nearly 100 samples of different natural sources (flowers, fruits, wood, and soil) from Kyoto, Japan, were utilized to isolate yeast strains able to ferment xylose. Among 72 isolates, 16 strains displayed a significant ethanol production (Tanimura et al. 2012).

A total of 133 xylose-utilizing yeast strains were isolated from 84 samples of decaying agricultural residues and soils collected in Thailand. Only eight strains were able to ferment xylose to ethanol, belonging to the species *Zygoascus hellenicus*, *Candida blankii*, and the novel species, *Candida saraburiensis* (Nitiyon et al. 2011). Another novel species proposed by these authors was *Candida prachuapensis*, a xylose-assimilating but not a xylose-fermenting yeast. Indeed, the number of species able to assimilate xylose is relatively high, but only a few have been reported to both assimilate and ferment D-xylose to ethanol (Hahn-Hagerdal et al. 2007).

Twenty-eight yeast strains capable of fermenting xylose to ethanol were isolated from the rectum of Murrah buffalo and Swamp buffalo in Thailand. The species included *Candida tropicalis*, *Candida parapsilosis*, *Candida mengyuniae*, *Sporopachydermia lactativora*, *Geotrichum* sp., and *Trichosporon asahii* (Lorliam et al. 2013).

About 770 yeast strains corresponding to nearly 78 species were isolated from Guatemalan passalids (Coleoptera). The xylose-fermenting yeasts *Scheffersomyces shehatae* and *S. stipitis* were the most common ascomycete yeasts identified. Regarding physiological profiles, the *Sugiyamaella* members all assimilate xylose and L- and D-arabinose, although most of them are not able to ferment xylose (Kurtzman et al. 2011; Urbina et al. 2013).

A total of 105 yeast strains were isolated from rotten wood samples from Baotianman Nature Reserve, China. Only 17 strains were xylose-fermenting and were identified as *Scheffersomyces insectosa*, *Scheffersomyces lignosus*, *Scheffersomyces segobiensis*, *S. stipitis*, *S. shehatae*, and *Spathaspora passalidarum.* Two novel species were characterized as *Scheffersomyces henanensis* (Ren et al. 2014).

In Brazil, the first study to identify new xylose-fermenting yeasts was carried out in the Brazilian Amazonian Forest by Cadete et al. (2012). A total of 224 yeast strains were isolated from rotting wood samples collected in the Amazonian forest, and six strains of *Spathaspora passalidarum*, two of *Scheffersomyces stipitis* (formerly known as *Pichia stipitis*), and representatives of five new species were identified as xylose-fermenting yeasts.

Forty-two yeast strains isolated from rotting wood of the Brazilian biodiversity were evaluated as xylose-fermenting and four of them were identified as new species of *Spathaspora*: *S. brasiliensis*, *S. roraimanensis*, *S. suhii*, and *S. xylofermentans* (Prompt 2012).

Yeast species from rotting wood of Atlantic rainforest ecosystems were investigated for xylose fermentation and xylanase production. Among 69 species identified, 15 possible new species were obtained. Two of them are new xylose-fermenting yeasts (*Sugiyamaella* sp. and *Sugiyamaella xylanicola*). Most of the xylanase-producing isolates belong to the new species *S. xylanicola*, also positive for xylose fermentation. *Scheffersomyces queiroziae* and *S. stipitis* were the main xylose-fermenting yeasts (Morais et al. 2013).

Martiniano et al. (2013) evaluated the ethanol production potential of four native strains of *S. shehatae* isolated from natural habitats in Brazilian forests, like bromeliads, mushroom, and *Euterpe* sp. The ethanol yield was higher in synthetic medium supplemented with xylose (0.44–0.48 g/g) than in acid hydrolysates (0.21–0.30 g/g). With immobilized cells of *S. shehatae* in a calcium alginate matrix, yield of 0.31 g/g was achieved in sugarcane bagasse hydrolysate (Antunes et al. 2016).

From rotting wood, 83 yeast colonies were obtained, and only 11 isolates were not able to produce ethanol from xylose. A range of 0.60–6.58 g/L ethanol was produced by the xylose-fermenting yeasts (Varize 2013).

From termites (*Nasutitermes* sp.) in the Amazonian habitat, a total of 54 yeast colonies were isolated, with 12 colonies able to assimilate xylose and one able to ferment it. The xylose-fermenting strain was identified as *Meyerozyma guilliermondii* and tested using sugarcane bagasse hydrolysates for ethanol production (Matos et al. 2014).

Among 350 isolates from natural sources such as sugarcane, bagasse, rhizosphere, and corn straw, 39 displayed fermentative activity of arabinose (56.4%) or xylose (38.5%) or both (5.1%). One of the yeast strains that showed both arabinose and xylose fermentation, isolated from sugarcane juice, was identified as *Meyerozyma guilliermondii* and tested in synthetic and bagasse hydrolysate for pentose fermentation (Martini 2014). The yeast grew in xylose, arabinose, and glucose at the same rate at the initial medium pH 5.5. At pH 4.5, the growth rate was slower in arabinose, and for the other carbon sources, there was no difference at different pH values (Martini et al. 2016).

Table 8.1 Some of the new yeast species isolated and described for xylose fermentation from various substrates in the last decade

Yeast species	Isolation	Reference
Enteroramus dimorphicus	Gut of passalid beetles in the United States	Suh et al. (2004)
Spathaspora passalidarum and *Candida jeffriesii*	Gut of passalid beetles in the United States and Panama	Nguyen et al. (2006)
Spathaspora arborariae	Rotting wood in the Atlantic Rain Forest and a Cerrado ecosystem in Brazil	Cadete et al. (2009)
Candida saraburiensis	Decaying agricultural residues and soils in Thailand	Nitiyon et al. (2011)
Wickerhamomyces xylosica and *Candida phayaonensis*	Soils of Thailand	Limtong et al. (2012)
Scheffersomyces henanensis	Rotten wood from the Baotianman Nature Reserve in Henan Province, Central China	Ren et al. (2014)

Although a large list of microorganisms that ferment pentoses was described in this chapter, not all the discoveries in this area of research was covered. In the last decade, novel species of yeasts have been described as xylose-fermenting and some examples are listed in Table 8.1.

8.3 Improving Nutritional and Cultural Conditions for Pentose Fermentations

For ethanol production from biomass containing high proportions of hemicellulose rich in pentoses, a series of steps comprising pretreatment, hydrolysis (chemical or enzymatic), and fermentation (Canilha et al. 2012; Laluce et al. 2012) are necessary. Both pretreatment and hydrolysis have effects on the fermentation step in which the candidate microorganism will convert the sugars released into ethanol. Among the chemical pretreatments of biomass, the acid hydrolysis is reported to be one of the commonest and oldest methods. The main disadvantages of this process are the need for neutralization before fermentation and the by-product formation, such as furans, furfural, carboxylic acids, acetic acid, and phenolic compounds. Some of these compounds can be inhibitors of microbial fermentation. Thus, a detoxification step is required following acid hydrolysis to remove inhibitory compounds and improve microbial fermentation (Palmqvist and Hahn-Hagerdal 2000).

Ethanol yield and productivity are strongly affected by the culture medium composition and other cultivation conditions such as nitrogen supplementation, aeration, temperature, pH, and sugar concentration (Jeffries 1986; du Preez 1994; Martiniano et al. 2013).

8.3.1 Medium Composition

The chemical composition of hydrolysates utilized for fermentation varies considerably concerning the plant species and hydrolysis conditions employed. The hemicellulose fraction of hardwoods and agricultural raw materials is richer in five-carbon sugars than softwood materials (Hayn et al. 1993). Xylose (12–16 g/L) is the main component in hemicellulose hydrolysate followed by arabinose (2–3 g/L) and glucose (0.7–7.6 g/L) (Mussato and Roberto 2004; Canilha et al. 2005; Martiniano et al. 2013; Milessi et al. 2013; Martini et al. 2016). Toxic compounds are also found in acid hydrolysates as furfural, 5-hydroxymethylfurfural, phenolics, weak acids, and others (Chandel et al. 2012).

Yeasts cultivated in a mix of sugars (hexoses and pentoses) may present an inhibition or a delay in the xylose utilization when glucose is present (Gong et al. 1999). Concentrations above 20 g/L of glucose repress the xylose utilization even when yeast species with high affinity by pentoses such as *P. stipitis* are utilized (Hahn-Hagerdal et al. 1991). However, the transport mechanism of pentose assimilation is usually activated when the glucose concentration is exhausted (Hou 2012). Some species, such as *S. passalidarum* and *C. shehatae*, displayed higher ethanol production from xylose than glucose (Mouro 2012).

Among the yeasts found in the Amazonia Forest by Cadete et al. (2012), glucose and D-xylose were initially metabolized at the same time, but the glucose consumption rate was clearly faster. The consumption rate of xylose was increased after glucose exhaustion. The sugar composition of the hydrolysate utilized in the experiments was 59 g/L xylose, 6.2 g/L glucose, and 6.4 g/L arabinose. It seemed, as already observed by Souto-Maior et al. (2009), that the application of glucose pulses enhanced xylose uptake under restricted mixed substrate concentrations.

Martini et al. (2016) reported that a strain of *Meyerozyma guilliermondii* consumed the sugars glucose, xylose, and arabinose, in this sequential order, with ethanol and xylitol yields of 0.14 and 0.19 g/g sugars, respectively, in non-detoxified hydrolysates (from sugarcane bagasse) with yeast extract supplementation.

With *P. stipitis,* the highest ethanol yields, 0.42 g/g and 0.47 g/g, were verified when the ratios between glucose and xylose initial concentrations were 20/5 and 24/1, respectively (Sanchez et al. 2002). In this work, *C. shehatae* and *P. tannophilus* were also evaluated, and for all three strains, hexoses and pentoses were sequentially consumed, glucose being the first.

The utilization of L-arabinose by yeasts has not received as much attention as xylose. A few studies have dealt with the arabinose metabolism and the culture conditions that drive the metabolism to the production of ethanol or arabitol. This may be partially due to the fact that arabinose utilization is considered less efficient than xylose. However, they share several control points (Almeida et al. 2011).

Fonseca et al. (2007a) studied the arabinose utilization by the yeasts *Candida arabinofermentans* and *Pichia guilliermondii* in aerobic batch cultures and compared to the glucose and xylose metabolisms. Both pentoses were essentially respired, and production of polyols—xylitol or arabitol—have occurred from xylose or arabinose only under oxygen limitation. This was the first study to demonstrate the oxygen dependence for metabolite and product formation in arabinose-assimilating yeasts.

The same yeast species above were studied to characterize the arabinose transport and the first steps of intracellular arabinose metabolism. The study revealed, at least, two transport systems and the catabolic pathway proposed for filamentous fungi also operated in the yeasts studied. The authors characterized for the first time a specific arabinose transporter (Fonseca et al. 2007b).

Besides the fact that the hemicellulosic hydrolysate is a multi-sugar substrate, the presence of inhibitors generated along the pretreatment and chemical hydrolysis of plant material influences considerably the microbial growth and fermentation. Three major groups include the inhibitory compounds: furaldehydes (furfural, 5-hydroxymethyl furfural-HMF), weak acids (acetic acid, formic acid, and levulinic acid), and phenolics (vanillin, syringaldehyde, and coniferyl aldehyde) (Palmqvist and Hahn-Hagerdal 2000; Mussato and Roberto 2004). The strategy of concentrating the hydrolysate by evaporation to obtain higher amounts of sugars, as it frequently occurs in the 1G generation ethanol industry, leads to the increase of the nonvolatile inhibitor concentration as well (Dehkhoda et al. 2009). On the other hand, furfural, HMF and acetic acid may have their concentrations decreased due to their volatility (Martiniano et al. 2013; Martini et al. 2016). The most studied inhibitory compound is acetic acid, which is present in hydrolysates at concentrations of 2–5 g/L (Tengborg et al. 2001; Martini et al. 2016). These different inhibitory compounds may also act synergistically affecting the microbial physiology (Parawira and Tekere 2011; Canilha et al. 2012).

The yeast *S. arborariae* in pure and mixed cultures with *S. cerevisiae* was negatively affected by the presence of toxic compounds like acetic acid, furfural, and HMF in the culture medium, decreasing biomass formation and ethanol productivity (Cunha-Pereira et al. 2011).

Some strategies are used to overcome the inhibitors in lignocellulose hydrolysates as physical/chemical detoxification, biological detoxification using microorganisms, enzymes or adaptation (evolutionary engineering) to toxic hydrolysate (Parawira and Tekere 2011).

Overliming, which consists in the addition of lime to increase the pH of the hydrolysate to precipitate toxic components, is one of the most used physicochemical treatments utilized to remove furans. Yet, other physicochemical applied treatments are ion exchange resins, neutralization, activated charcoal and extraction with organic solvents (Canilha et al. 2012). Nonetheless, loss of sugars is reported when neutralization and activated charcoal are utilized as detoxifiers, in spite of the significant decrease in inhibitor concentrations (Martiniano et al. 2013; Martini et al. 2016).

The biological detoxification consists in the usage of specific enzymes or microorganisms that act on the inhibitory compounds and promote changes in their chemical constitution (Parawira and Tekere 2011; Canilha et al. 2012). The use of enzymes such as laccases and peroxidases has been found to be effective to remove phenolic compounds (Chandel et al. 2011). Microorganisms can naturally assimilate inhibitory compounds such as acetic acid, furfural, and guaiacol. Yeast species as *Issatchenkia occidentalis, Iris occidentalis, M. guilliermondii*, and *S. cerevisiae* have been reported to be hydrolysate detoxifiers (Talebnia and Taherzadeh 2006; Hou-Rui et al. 2009; Martini et al. 2016). The advantages of using a biological treatment instead of a physicochemical treatment include fewer side reactions, fewer energy requirements, feasibility, and environmental friendliness. On the other hand, it is a long-time consuming process (Yang and Wyman 2008; Chandel et al. 2011; Parawira and Tekere 2011).

In addition, a particular microorganism may be adapted to inhibitors in a long-term process under selective pressure, by evolutionary engineering. This process is a promising alternative to the physicochemical detoxification step (Canilha et al. 2012). Under laboratory conditions, variants of the cell population with a selective advantage can take over exponentially the initial dominating cells. A great advantage to using this process is the fact that the evolution of a fermenting microorganism may be carried out in a whole hydrolysate containing a cocktail of inhibitors. Adaptation to lignocellulosic hydrolysates resulted in increased fermentation rates and ethanol yields (Parawira and Tekere 2011).

Wisselink et al. (2009) first described a strategy utilized to get an evolved *S. cerevisiae* strain able to ferment a medium containing glucose, xylose, and arabinose with a high ethanol yield (0.43 g/g total sugar) without formation of side products as xylitol and arabinitol. They utilized a schedule consisting of repeated batch cultivation with repeated cycles of consecutive growth in media with different compositions of glucose, xylose, and arabinose.

A recombinant industrial *S. cerevisiae* strain carrying the heterologous genes for xylose and arabinose utilization pathways integrated into the genome was evolved to improve simultaneous conversion of both pentoses to ethanol under aerobic and anaerobic conditions. Arabinose was totally converted to arabitol, but ethanol, xylitol, and glycerol were produced exclusively from xylose under anaerobic conditions. This work pioneered the characterization of the molecular mechanisms for improved mixed-pentose utilization obtained by evolutionary engineering of a recombinant *S. cerevisiae* strain (Sanchez et al. 2010).

To improve the inhibitor tolerance in a metabolically engineered xylose-utilizing strain of *S. cerevisiae*, a long-term adaptation was employed in repetitive batch cultures using a cocktail of 12 different inhibitors and a long term chemostat adaptation using spruce hydrolysate (Koppram et al. 2012). The yeast was evolved for 429 generations in repetitive batch cultures, increasing the growth rate from 0.18 to 0.33 h^{-1}, and for 97 generations in chemostat cultivation, decreasing the lag phase from 48 to 24 h. Three evolved strains were isolated.

Wallace-Salinas and Gorwa-Grauslund (2013) verified that an evolved isolate (ISO12) from *S. cerevisiae* showed a superior phenotype than the parental strain ER when both temperature stress and inhibition by hydrolysate-derived compounds were applied together.

Smith et al. (2014) reported that the sequential application of random mutagenesis followed by continuous culture under simultaneous selective pressure from inhibitors and xylose as the primary carbon source resulted in a strain of

S. cerevisiae that exhibited a shorter lag phase. It also removed completely HMF, furfural, and acetic acid from the fermentation broth, with concomitant ethanol production from xylose.

8.3.2 Nitrogen Supplementation

An adequate nitrogen supplementation is essential for cell growth and to reduce/avoid the negative effects caused by the presence of inhibitors (Silva et al. 2014). The most common sources of nitrogen are ammonium sulfate, urea, peptone, malt extract, and yeast extract (Ferreira et al. 2011; Cadete et al. 2012; Martiniano et al. 2013; Antunes et al. 2014; Silva et al. 2014; Martini et al. 2016).

Four different formulations of fermentation media were evaluated by Antunes et al. (2014) for ethanol production by *S. shehatae* using bagasse hydrolysate. The most adequate formulation to produce better results for ethanol yield and productivity was 5 g/L ammonium sulfate, 3 g/L yeast extract, and 3 g/L malt extract. The use of yeast extract and urea was also reported by Ferreira et al. (2011).

However, the products used routinely such as yeast extract and peptone as nitrogen source are expensive and contribute to almost 50% of the cost of ethanol production (Ananda et al. 2011). Some alternative nitrogen sources have emerged. For instance, rice bran, a residue of rice milling which contains amino acids, vitamins, proteins, and minerals seems to be a good substitute (Miller and Churchill 1986; Milessi et al. 2013; Silva et al. 2014). In fact, the ethanol yield from bagasse hydrolysates by *S. stipitis* was higher when rice bran extract was utilized instead of other nitrogen supplements (Milessi et al. 2013).

The findings of Silva et al. (2014) indicated that the best results depended on the yeast strain and the concomitant addition of an inorganic nitrogen source, as ammonium sulfate. For *S. stipitis*, the hydrolysate must be supplemented with peptone and yeast extract. For *C. guilliermondii*, rice bran extract presented the best results, but for both yeasts, ammonium sulfate is also required as supplementation. In a synthetic medium, ammonium salts increased the ethanol productivity of *P. stipitis* (Agbogbo and Wenger 2006). Corn steep liquor was also used successfully as a nitrogen source for *P. stipitis* (Amartey and Jeffries 1994) and engineered yeast strains (Sarks et al. 2014) in fermentation.

8.3.3 Oxygen

Among the operational conditions, the oxygen availability is considered the most important factor affecting pentose fermentation by yeasts, since it determines the partitioning of the carbon flow substrate between growth and product formation (du-Preez 1994) and plays an important role in cell growth, redox balance, and functioning of the mitochondria (Skoog and Hahn-Hagerdal 1990). Thus, the level of oxygen can drive the product formation during fermentation of xylose. In aerobic conditions, the microbial metabolism is deviated for cell production, decreasing the xylitol production. Under oxygen-limiting conditions, xylitol accumulation occurs (Náhlik et al. 2003). However, under anaerobic conditions, xylitol production is negligible (Vandeska et al. 1995).

In the early 1980s, *C. shehatae* and *P. stipitis* were found to be dependent on aeration to enhance the ethanol productivity (Schneider et al. 1981; du Preez et al. 1984). Since then, many studies have pointed the requirement of oxygen to obtain better results in xylose fermentation. Indeed, microaerobic conditions are the most suitable operating conditions for this specific fermentation (Grootjen et al. 1990; Agbogbo and Coward-Kelly 2008; Silva et al. 2011; Bellido et al. 2013; Su et al. 2015). Nevertheless, Krahulec et al. (2012) showed that a low level of oxygen is not a requirement to achieve high ethanol yields from xylose by *P. stipitis.*

The establishment of an adequate oxygen level is thus of great importance to obtain an efficient process with high values of conversion and productivity. In this sense, Silva et al. (2012) verified the influence of the oxygen transfer in xylose conversion to ethanol by *P. stipitis*. The authors reported that the oxygen availability to the medium was essential to guarantee elevated ethanol production by this yeast, but this variable must be carefully controlled since the excess affects the carbon source metabolism, favoring the cell growth in detriment to the ethanol formation. In fact, moderate oxygen reduces xylitol formation and increases ethanol production by reoxidizing NADH in the cytosol through respiration metabolism. However, it is complex and uneconomical to perform the oxygen control in industrial ethanol production. In this way, respiration-independent *S. cerevisiae* strains might be more beneficial for industrial ethanol production from xylose (Peng et al. 2012).

In fact, the issue of oxygen requirement by *P. stipitis* could be better understood in the light of the genome sequencing of this species. The strain sequenced (CBS 6054) was isolated from insect larvae, similarly to other strains of *P. stipitis* that were also isolated from the guts of wood-inhabiting passalid beetles (Suh et al. 2003), suggesting that this yeast species is a common inhabitant of oxygen-limited environments rich in partially digested wood. The high efficiency of xylose fermentation demonstrated by *P. stipitis* under microaerobic conditions is due to a well-handled regulation of redox balance. The abundance of genes for NADP(H) oxidoreductase reactions indicates that diverse strategies for balancing NAD and NADP-specific cofactors are present in *P. stipitis* (Jeffries et al. 2007).

Galafassi et al. (2011) utilized a strategy based on the management of the aeration level to obtain ethanol from synthetic lignocellulosic hydrolysates by strains of *Dekkera/Brettanomyces* yeasts. In the first step, aerobic conditions resulted in fast biomass production from mixtures of hexoses and pentoses, and in the second step, under oxygen limitation, the ethanol was produced. Ethanol yields ranging from 0.2 to 0.3 g/g were obtained, with consumption of glucose, xylose, and arabinose. Despite the low specific growth rate and ethanol production, a strain of *Dekkera bruxellensis* was able to grow and ferment sugarcane bagasse hydrolysates (Codato 2013).

8.3.4 pH

The influence of the pH of the culture medium may vary among yeast strains, but in all cases, it affects cell growth, cell membrane permeability, and the solubility of components of the culture medium. For *C. shehatae*, an adequate pH for growth was between 3.5 and 4.5 (du Preez et al. 1986). For ethanol production, the best pH was 4.5 (Sanchez et al. 1997, 2004). Concerning *P. stipitis,* the optimum range for growth and fermentation in the presence of xylose was 4–7 at 25 °C (Slininger et al. 1990).

As mentioned above, acetic acid is a prominent inhibitor of xylose uptake, resulting in gradual acidification during fermentation. The effective control of pH reduced the inhibition by the acetic acid and enhanced specific xylose uptake rates as well as ethanol yields (Novy et al. 2013). In low pH, acetic acid becomes liposoluble and is able to diffuse through the plasmatic membrane. Inside the cell, it dissociates due to the neutral pH and is accumulated into the cytoplasm. The protons, released into the cytoplasm, result in the decrease in the cell pH causing inhibition of the cell activity and death (Chandel et al. 2012).

In pH values below 4.5, the consumption of the carbon sources by *C. guilliermondii* was strongly inhibited in bagasse hydrolysates. Morphological changes in the cell membrane, disturbances in the active transport through the membrane due to the pH alterations, and the competition between sugar and acetic acid for active sites are the main causes of the inhibition of sugar consumption in low pH (Felipe et al. 1997).

8.3.5 Temperature

The temperature effects on ethanol and xylitol formation are different. The polyol yield is less affected by variations in temperature than ethanol yield, and this fact may drive the production for one of the products when desired (Sanchez et al. 2004). The production of xylitol decreased substantially, and the ethanol production did not change when the temperature was raised from 30 to 37 °C during xylose fermentation by *P. tannophilus* (Barbosa et al. 1990). This suggests that oxidative phosphorylation becomes uncoupled at higher temperature. For *P. stipitis*, the optimal fermentation temperature was 25–26 °C for xylose and 30 °C for glucose (Jeffries 2008).

About 83% of xylose was consumed by *P. tannophilus* to produce xylitol at a temperature range from 25 to 30 °C; however, alcohol and biomass production were significant outside this temperature range (Converti et al. 2001).

The use of *Kluyveromyces marxianus* for potential application in 2G ethanol has been disseminated (Signori et al. 2014; Nitiyon et al. 2016). This yeast is able to produce ethanol at high temperatures, assimilate pentoses and hexoses, and weak glucose repression in sucrose-based substrates (Limtong et al. 2007; Lertwattanasakul et al. 2011). Two *K. marxianus* strains were able to ferment xylose at 37 °C, but *S. stipitis* was not. *K. marxianus* is superior to *S. stipitis* regarding thermotolerance but the ethanol productivity from xylose is higher with the last yeast species (Nitiyon et al. 2016).

8.4 Engineering Yeasts to Achieve Better Performance in Pentose Fermentation

As discussed in previous sections of this chapter, the challenges for using yeasts during the ethanol production from lignocellulosic substrates are multiple and diverse. The nature of the culture medium, the presence of inhibitors generated during the pretreatment and hydrolysis steps, the formation of products as ethanol in high titer, osmotic stress, and other toxic effects are important characteristics of the 2G generation bioethanol process (Almeida et al. 2011). A single yeast capable of overcoming this combination of environmental factors and with a genetic background that allows efficient conversion of pentoses to ethanol has not been easy to found.

The species *S. cerevisiae* is the most extensively used yeast in fermentation processes with sugar- and starch-based raw materials, and it is greatly adapted to the industrial context and its peculiarities. However, it is unable to ferment pentoses (Hahn-Hagerdal et al. 2007). To make the process feasible, it is crucial to have an organism that gathers most of the required characteristics for this particular fermentation. Natural xylose-fermenting yeasts may be a reservoir of genes to be expressed in *S. cerevisiae* by the usage of genetic engineering. In this context, great effort has been concentrated on the construction of recombinant yeast strains with increased xylose fermentation capacity over the past decades.

Many approaches have been utilized to engineer yeasts for xylose metabolism as described in Jeffries (2006, 2008), Hahn-Hagerdal et al. (2007), Almeida et al. (2011), Laluce et al. (2012), and Harner et al. (2015). Strategies of modeling, flux analysis, expression analysis, targeted deletion, random mutagenesis, genome shuffling, and altered expression of key genes have been employed in laboratory and industrial yeast strains.

Richard et al. (2003) combined the genes from different fungi to construct a strain able to utilize arabinose. The genes *XYL1* and *XYL2* from *P. stipitis*, *XKS1* from *S. cerevisiae*, and *lad1* and *lxr1* from *Trichoderma reesei* were inserted in plasmids or integrated into the chromosomes of *S. cerevisiae.* The resulting strain showed activities of all the enzymes of the pathways and was able to assimilate arabinose, but at a very low rate. The imbalance of redox factors and the arabinose transport into the cell are possible limitation factors for arabinose fermentation. For the first time, those authors demonstrated the production of ethanol from arabinose in anaerobic conditions.

Wisselink et al. (2007) also reported a case of fast and efficient anaerobic alcoholic fermentation of arabinose by an engineered *S. cerevisiae* strain. The strain was constructed by combining the expression of the structural genes of *Lactobacillus plantarum* for arabinose utilization, the overexpression of *S. cerevisiae* genes encoding the enzymes of the nonoxidative pentose phosphate pathway, and extensive evolutionary engineering. A relatively high ethanol yield (0.43 g/g) was achieved by growing the engineered yeast on arabinose as the sole carbon source anaerobically.

Simultaneous cellobiose and xylose fermentation were also achieved by engineering strains of *S. cerevisiae* (Ha et al. 2011). The authors expressed both wild-type xylose reductase and a mutant xylose reductase in *S. cerevisiae,* along with xylitol dehydrogenase and xylulokinase in order to construct a functional xylose metabolic pathway in this yeast. They utilized the strategy of evolutionary engineering to obtain a strain that rapidly consumed xylose and produced consistent ethanol yields (0.31–0.32 g/g). To increase the efficiency of xylose consumption, these authors also introduced genes coding for a cellodextrin transporter and a β-glucosidase. The strain containing all genes (DA24-16) was capable of consuming cellobiose and xylose simultaneously.

In the study by Zhang and Geng (2012), *S. cerevisiae* and *P. stipitis* were used as parents for recombinant yeast strain construction. After two rounds of genome shuffling and screening, a recombinant (ScF2) was obtained with the ability to ferment high concentration of xylose (100–250 g/L) into ethanol faster than *P. stipitis*.

A novel xylose-fermenting yeast strain, FSC1, was developed by intergeneric hybridization between *S. cerevisiae* and *Candida intermedia* mutants using a protoplast fusion technique. An ethanol yield of 0.38 g/g in a mix of substrates composed of glucose and xylose was achieved. The successive chemical mutation resulted in higher ethanol yield of 0.42 g/g substrate (Kahar and Tanaka 2014).

The genetic modification in laboratory strains of *S. cerevisiae* is relatively easier, especially because they are usually haploid and with a well-known genetic composition. However, industrial strains are diploid, polyploid, or aneuploid, with multiple copies of chromosomes and few auxotrophic marks. The genetic engineering of industrial strains have been limited to the introduction of the initial xylose and arabinose utilization pathways in *S. cerevisiae* (Hahn-Hagerdal et al. 2007), such as the enzymes xylose reductase (XR) and xylitol dehydrogenase (XDH) from *P. stipitis*.

To change the preference of XR for NADPH in anaerobic conditions due to low availability of NAD+ for XDH (Fig. 8.1), which would result in xylitol formation and less ethanol production, Bengtsson et al. (2009) utilized site-directed mutagenesis to direct the preference towards NADH in *P. stipitis*. In sequence, a *S. cerevisiae* was constructed harboring the mutated XR. The fermentation yields presented by the recombinant yeast were 0.39 g/g for ethanol and 0.05 g/g for xylitol, in continuous fermentation at a dilution rate of 0.12 h^{-1}, with a mixture of glucose and xylose (10 g/L each), anaerobically.

Employing the recombinant *S. cerevisiae* described above, Runquist et al. (2009) demonstrated that similar product yields were obtained both for xylose and glucose. However the ethanol productivity from xylose was lower. Cofactor imbalance in the initial two steps of xylose utilization may be responsible for the decrease in ethanol productivity.

A redox balancing strategy to an efficient xylose fermentation with simultaneous consumption of acetic acid by *S. cerevisiae* was proposed by Wei et al. (2013). The authors combined a NADH-consuming acetate consumption pathway and a NADH-producing xylose utilization pathway and the engineered yeast co-consumed xylose and the fermentation inhibitor acetate under anaerobic conditions to produce ethanol.

All the studies described in this section show that genetic manipulation might be successful in gathering robustness against all the inhibitors generated along the hydrolysis processes and higher fermentation performance in a single strain. In this context, the genomic analysis of *P. stipitis* and the newly yeast strains isolated as *S. arborariae* may reveal interesting genes for biotechnological traits (Jeffries et al. 2007; Lobo et al. 2014).

8.5 Technological Scenarios for 1G and 2G Ethanol

Technical and economical feasibility must yet to be proven for the technology of 2G ethanol to be fully absorbed by industry. This situation was not different for the ethanol produced from hexose-based substrates since its implementation in the 1970s in Brazil, which is now the most competitive producer of bioethanol from sugarcane in the world. New technologies were developed and transferred to distilleries to increase the industrial efficiency (Amorim et al. 2011). What can we learn from this successful technology that could be employed in the pentose fermentation to reach industrial performance? The characteristics of the industrial ethanol production from sugarcane musts in Brazil were compared to the pentose fermentation in the light of the fermentation parameters (Table 8.2). The comparison is undoubtedly unfair for the pentoses because the industrial microorganism used for ethanol production from sugarcane musts is *S. cerevisiae,* which has no competitor with similar qualities. For pentose fermentation, the most important yeast species is *P. stipitis*, but the number of species is increasing. Furthermore, there are some recombinant strains of *S. cerevisiae* that are promising.

The main peculiarities of the Brazilian ethanol production process from sugarcane are laid on the facts that the process is not carried out under aseptic conditions, and the yeast cells are constantly recycled along the fermentative cycles, around 400–600 cycles (Amorim et al. 2011). Sarks et al. (2014) utilized 24-h high cell density fermentations along with cell recycling to solve the slow/incomplete xylose fermentation, which is considered critical for lignocellulosic fermentation. Not all the strains tested were capable of efficiently performing the process. The xylose consumption depended on the initial cell concentration and the nutrient supplementation. The main cause of reduced xylose consumption with subsequent cycles is the decrease of specific xylose consumption rate rather than decreased viable cell mass. This approach for pentose fermentation is an advance to reduce fermentation time and improve ethanol productivity.

Table 8.2 Characteristics of the industrial ethanol production from sugarcane musts in Brazil compared to the pentose fermentation from hemicellulosic substrates

Process	Initial cell concentration	Substrate	Fermentation batch time	Fermentation process	Oxygen requirement	Ethanol yield	Ethanol productivity
Ethanol from sugar cane (*S. cerevisiae*)	High yeast cell densities (10–15% w/v)[a]	Sugarcane juice and or molasses (sucrose-based substrates); no previous treatment required; 20–25% (w/v)[a]	6–12 h[a]	Fed-batch or continuous with cell recycle[a]	No	92–93% (or 0.47 g/g)[a]	~2–3 g/L h[b]
Ethanol from hemicellulose (*P. stipitis*)	Low yeast cell densities (0.01–10 g/L)[c,d,e,f]	Multiple-sugar substrate (glucose, arabinose and xylose)[g]; pretreatment and hydrolysis required; 50 g/L xylose or less[h,i]	Variable, from 48 h[j] to 192 h[k]	Batch without cell recycle	Yes (1.5–5 mmol/L h)[e,j]	0.4–0.47 g/g[j,d]	0.1–0.9 g/L h[e,f,i]
Observation for ethanol from hemicellulose	This parameter has not received much attention. It affects fermentation rate, ethanol yield and concentration[d]	Need to overcome the glucose repression[g]. Discrepancies about the inhibitory effects of high xylose concentration	Lower consumption rate than *S. cerevisiae*; sequential consumption of glucose and xylose[l]	Recent work[m] has studied a system using high cell density with cell recycle with hydrolysates of corn stover	It is considered the most important factor that drives pentose fermentation[h,j]	Yields are comparable	Very low productivity due to the slow fermentation

[a]Amorim et al. (2011)
[b]Andrietta et al. (2011)
[c]Sreenath and Jeffries (2000)
[d]Sanchez et al. (2002)
[e]Agbogbo et al. (2007)
[f]Milessi et al. (2015)
[g]Hahn-Hagerdal et al. (2007)
[h]Silva et al. (2010)
[i]Cadete et al. (2009)
[j]Su et al. (2015)
[k]Agbogbo et al. (2007)
[l]Agobgo and Coward-Kelly (2008)
[m]Sarks et al. (2014)

Regarding the microbial contamination issue, as far as we know the dominance of the pentose-fermenting yeast in relation to other yeasts and bacteria contaminating the substrate has not received attention yet. This is a real challenge to be faced by the selected yeast, whichever it will be (Almeida et al. 2011).

8.6 Conclusions and Perspectives

Many obstacles and challenges involve the full development of 2G ethanol. The steps of pretreatment and chemical hydrolysis of the plant biomass generate undesirable compounds that affect the fermentation. Although the enzymatic hydrolysis is an alternative to avoid the formation of these compounds, it is still a high-cost process. Regarding the fermentation step, which is the core of this chapter, the challenges to be overcome are quite large. Attempts to find naturally occurring yeast strains that can ferment pentoses have been successful, and *P. stipitis* has been one of the most important yeast species utilized in fermentations carried out in synthetic media or hydrolysates from varied plant sources. The parameters that drive the fermentation by this yeast are not completely known, with the exception of the requirement for oxygen. It is noteworthy that the influence of initial cell concentration has not gained much attention yet. Even considering the specificities of the process and of the substrates, there is much to be learned from the experience of 1G ethanol. High cell densities are obtained in this process by recycling the cells after each fermentative cycle, which guarantees high ethanol yield and shorter fermentation times. As far as we are concerned, only sparse experiments have dealed with cell recycling in pentose fermentation. Genome sequences of newly isolated strains may reveal interesting genes related to biotechnological traits and capable to facilitate genetic engineering of *S. cerevisiae*, thus allowing the production of a robust strain from this species able to ferment pentoses with higher efficiency. Despite the challenges yet to be faced, research must go towards the direction of searching for yeast with analogous distinct features as the ones exhibited by *S. cerevisiae* for the 1G ethanol technology.

Acknowledgments The authors wish to thank CNPq (Conselho Nacional de Desenvolvimento Científico e Tecnológico) and FAPESP (Fundação de Amparo à Pesquisa do Estado de São Paulo) for the financial support to INCT (Instituto Nacional de Ciência e Tecnologia do Bioetanol).

References

Agbogbo FK, Coward-Kelly G (2008) Cellulosic ethanol production using the naturally occurring xylose-fermenting yeast, *Pichia stipitis*. Biotechnol Lett 30:1515–1524

Agbogbo FK, Coward-Kelly G, Torry-Smith M, Wenger K, Jeffries TW (2007) The effect of initial cell concentration on xylose fermentation by *Pichia stipitis*. Appl Biochem Biotechnol 136:653–662

Agbogbo FK, Wenger KS (2006) Effect of pre treatmentchemicals on xylose fermentation by *P. stipitis*. Biotechnol Lett 28(24):2065–2069

Almeida JRM, Runquist D, Nogué VS, Lidén G, Gorwa-Grauslund MF (2011) Stress-related challenges in pentose fermentation to ethanol by the yeast *Saccharomyces cerevisiae*. Biotechnol J 6:286–299

Amartey SA, Jeffries TW (1994) Comparison of corn steep liquor with other nutrients in the fermentation of D-xylose by *Pichia stipitis* CBS6054. Biotechnol Lett 16(2):211–214

Amorim HV, Lopes ML, Oliveira JVC, Buckeridge MS, Goldman GH (2011) Scientific challenges of bioethanol production in Brazil. Appl Microbiol Biotechnol 91:1267–1275

Ananda N, Vadlani PV, Madl RL (2011) Rice bran is an effective substitute for yeast extract in ethanol fermentation. J Biobased Mater Bioenergy 5(1):70–74

Andrietta MGS, Andrietta SR, Stupiello ENA (2011) Bioethanol—what has Brazil learned about yeasts inhabiting the ethanol production processes from sugar cane? In: MAS B (ed) Biofuel production—recent developments and prospects. InTech, Croatia, pp 67–84

Antunes FAF, Chandel AK, Milessi TSS, Santos JC, Rosa CA, Silva SS (2014) Bioethanol production from sugarcane bagasse by a novel Brazilian pentose fermenting yeast *Scheffersomyces shehatae* UFMG-HM 52.2: evaluation of fermentation medium. Int J Chem Eng. doi:10.1155/2014/180681

Antunes FAF, Santos JC, Chandel AK, Milessi TSS, Peres GFD, Silva SS (2016) Hemicellulosic ethanol production by immobilized wild Brazilian yeast *Scheffersomyces shehatae* UFMG-HM 52.2: effects of cell concentration and stirring rate. Curr Microbiol 72:133–138

Barbosa MFS, Lee H, Schneider H, Forsberg CW (1990) Temperature mediated changes of D-xylose metabolism in the yeast *Pachysolen tannophilus*. FEMS Microbiol Lett 72:35–40

Barnett JA (1976) The utilization of sugars by yeasts. Adv Carbohydr Chem Biochem 32:125–134

Bellido C, González-Benito G, Coca M, Lucas S, García-Cubero MT (2013) Influence of aeration on bioethanol production from ozonized wheat straw hydrolysates using *Pichia stipitis*. Bioresour Technol 133:51–58

Bengtsson O, Hahn-Hagerdal B, Gorwa-Grauslund MF (2009) Xylose reductase from *Pichia stipitis* with altered coenzyme preference improves ethanolic xylose fermentation by recombinant *Saccharomyces cerevisiae*. Biotechnol Biofuels 2:9–19

Botha A (2006) Yeasts in soil. In: Rosa CA, Péter G (eds) Biodiversity and ecophysiology of yeasts. Springer, Heidelberger, pp 221–240

Boundy-Mills K (2006) Methods for investigating yeast biodiversity. In: Rosa CA, Péter G (eds) Biodiversity and ecophysiology of yeasts. Springer, Heidelberger, pp 67–100

Buckeridge MS, Santos WD, Tiné MS, De Souza AP (2015) The cell wall architecture of sugarcane and its implications to cell wall recalcitrance. In: Lam E, Carrer H, Silva JA (eds) Compendium of bioenergy plants: sugarcane. CRC Press—Taylor & Francis Group, Boca Raton, FL, pp 31–50

Cadete RM, Melo MA, Dussán KJ, Rodrigues RCLB, Silva SS, Zilli JE, Vital MJS, Gomes FCO, Lachance MA, Rosa CA (2012) Diversity and physiological characterization of D-xylose fermenting yeasts isolated from the Brazilian Amazonian Forest. PLoS One 7(8):e43135. doi:10.371/journal.pone.0043135

Cadete RM, Santos RO, Melo MA, Mouro A, Gonçalves DL, Stambuk BU, Gomes FCO, Lachance MA, Rosa CA (2009) *Spathaspora arborariae* sp. nov., a D-xylose-fermenting yeast species isolated from rotting wood in Brazil. FEMS Yeast Res 9:1338–1342

Canilha L, Carvalho W, Silva JBA (2005) Influence of medium composition on xylitol bioproduction from wheat straw hemicellulosic hydrolysate. World J Microbiol Biotechnol 21:1087–1093

Canilha L, Chandel AK, Milessi TSS, Antunes FAF, Freitas WLC, Felipe MGA, Silva SS (2012) Bioconversion of sugarcane biomass into ethanol: an overview about composition, pretreatment methods, detoxification of hydrolysates, enzymatic saccharification, and ethanol fermentation. J Biomed Biotechnol. doi:10.1155/2012/989572

Carpita NC, Gibeaut DM (1993) Structural models of primary cell walls in flowering plants: consistency of molecular structure with the physical properties of the walls during growth. Plant J 3(1):1–30

Chandel AK, Chandrasekhar G, Radhika K, Ravinder R, Ravindra P (2011) Bioconversion of pentose sugars into ethanol: a review and future directions. Biotechnol Mol Biol Rev 6:8–20

Chandel AK, Silva SS, Singh OV (2012) Detoxification of lignocelluloses hydrolysates: biochemical and metabolic engineering toward white biotechnology. Bioenergy Res 6(1):388–401

Codato CB (2013) Produção de etanol de segunda geração por Dekkera bruxellensis a partir de hidrolisado de bagaço de cana-de-açúcar. Dissertation, Universidade Federal de São Carlos

Converti A, Perego P, Domínguez JM, Silva SS (2001) Effect of temperature on the microaerophilic metabolism of *Pachysolen tannophylus*. Enzyme Microb Technol 28:339–345

Cunha-Pereira F, Hickert LR, Sehnem NT, Souza-Cruz PB, Rosa CA, Ayub MAZ (2011) Conversion of sugars present in rice hull hydrolysates into ethanol by *Spathaspora arborariae, Saccharomyces cerevisiae*, and their co-fermentations. Bioresour Technol 102:4218–4225

Deak T (2009) Ecology and biodiversity of yeasts with potential value in biotechnology. In: Satyanarayana T, Kunze G (eds) Yeast biotechnology: diversity and applications. Springer, Heidelberger, p 151

Dehkhoda A, Brandberg T, Taherzadeh M (2009) Comparison of vacuum and high pressure evaporated wood hydrolyzate for ethanol production by repeated fed-batch using flocculating *Saccharomyces cerevisiae*. BioRes 4(1):309–320

De Souza AP, Grandis A, Leite DCC, Buckeridge MS (2014) Sugarcane as a bioenergy source: history, performance, and perspectives for second-generation bioethanol. Bioenerg Res 7:24–35

De Souza AP, Kamei CLA, Torres AF, Pattathil S, Hahn MG, Trindade LM, Buckeridge MS (2015) How cell wall complexity influences saccharification efficiency in *Miscanthus sinensis*. J Exp Bot 66:4351–4365

De Souza AP, Leite DCC, Pattathil S, Hahn MG, Buckeridge MS (2013) Composition and structure of sugarcane cell wall polysaccharides: implications for second-generation bioethanol production. Bioenerg Res 6:564–579

du Preez JC (1994) Process parameters and environmental factors affecting D-xylose fermentation by yeasts. Enzyme Microb Technol 16:944–952

du Preez JC, Bosh M, Prior BA (1986) Xylose fermentation by *Candida shehatae* and *Pichia stipitis*: effects of pH, temperature and substrate concentration. Enzyme Microb Technol 8(6):360–364

du Preez JC, Prior BA, Monteiro AMT (1984) The effect of aeration on xylose fermentation by *Candida shehatae* and *Pachysolen tannophilus*. Appl Microbiol Biotechnol 19:261–266

du Preez JC, van der Walt JP (1983) Fermentation of D-xylose to ethanol by a strain of *Candida shehatae*. Biotechnol Lett 5(5):357–362

Fan TWM, Lane AN, Shenkar M, Bartley JP, Crowley D, Higashi RM (2001) Comprehensive chemical profiling of gramineous plant root exudates using high-resolution NMR and MS. Phytochemistry 57:209–221

Felipe MGA, Vitolo M, Mancilha IM, Silva SS (1997) Fermentation of sugar cane bagasse hemicellulosic hydrolysate for xylitol production: effect of pH. Biomass Bioenerg 13(1/2):11–14

Ferreira AD, Mussatto SI, Cadete RM, Rosa CA, Silva SS (2011) Ethanol production by a new-pentose fermenting yeast strain, *Scheffersomyces stipitis* UFMG-IMH 43.2, isolated from the Brazilian forest. Yeast 28:547–554

Fonseca C, Romão R, Sousa HR, Hahn-Hagerdal B, Spencer-Martins I (2007b) L-Arabinose transport and catabolism in yeast. FEBS J 274:3589–3600

Fonseca C, Spencer-Martins I, Hahn-Hagerdal B (2007a) L-Arabinose metabolism in *Candida arabinofermentans* PYCC 5603^T and *Pichia guilliermondii* PYCC 3012: influence of sugar and oxygen on product formation. Appl Microbiol Biotechnol 75:303–310

Galafassi S, Merico A, Francesca P, Hellborg L, Molinari F, Piskur J (2011) *Dekkera/Brettanomyces* yeasts for ethanol production from renewable sources under oxygen-limited and low-pH conditions. J Ind Microbiol Biotechnol 38(8):1079–1088

Gong CS, Cao NJ, Du J, Tsao GT (1999) Ethanol production from renewable resources. Adv Biochem Eng Biotechnol 65:207–241

Grootjen DRJ, van der Lans RGJM, Luyben KCA (1990) Effects of the aeration rate on the fermentation of glucose and xylose by *Pichia stipitis* CBS 5773. Enzyme Microb Technol 12:20–23

Ha SJ, Galazka JM, Kim SR, Choi JH, Yang X, Seo JH, Glass LN, Cate JHD, Jin YS (2011) Engineered *Saccharomyces cerevisiae* capable of simultaneous cellobiose and xylose fermentation. PNAS 108(2):504–509

Hahn-Hagerdal B, Karhumaa K, Fonseca C, Spencer-Martins I, Gorwa-Grauslund MF (2007) Towards industrial pentose-fermenting yeast strains. Appl Microbiol Biotechnol 74:937–953

Hahn-Hagerdal B, Linden T, Senac T, Skoog K (1991) Ethanolic fermentation of pentoses in lignocellulose hydrolysate. Appl Biochem Biotechnol 28-29(1):131–144

Hahn-Hagerdal B, Pamment N (2004) Microbial pentose metabolism. Appl Biochem Biotechnol 113:1207–1209

Harner NK, Wen X, Bajwa PK, Austin GD, Ho CY, Habash MB, Trevors JT, Lee H (2015) Genetic improvement of native xylose-fermenting yeasts for ethanol production. J Ind Microbiol Biotechnol 42:1–20

Hayn M, Steiner W, Klinger R, Steinmuller H, Sinner M, Esterbauer H (1993) Basic research and pilot studies on the enzymatic conversion of lignocellulosics. In: Saddler JN (ed) Bioconversion of forest and agricultural plant residues. CAB International, Wallingford, UK, pp 33–72

Hendriks AT, Zeeman G (2009) Pretreatments to enhance the digestibility of lignocellulosic biomass. Bioresour Technol 100(1):10–18

Hou X (2012) Anaerobic xylose fermentation by *Spathaspora passalidarum*. Appl Microbiol Biotechnol 94:205–214

Hou-Rui Z, Xiang-Xiang Q, Silva SS, Sarrouh BF, Ai-Hua C, Yu-Heng Z, Ke J, Qiu X (2009) Novel isolates for biological detoxification of lignocellulosic hydrolysate. Appl Biochem Biotechnol 152(2):199–212

Jeffries TW (1986) Effects of culture conditions on the fermentation of xylose to ethanol by *Candida shehatae*. Biotechnol Bioeng 15:149–166

Jeffries TW (2006) Engineering yeasts for xylose metabolism. Curr Opin Biotechnol 17:320–326

Jeffries TW (2008) Engineering the *Pichia stipitis* genome for fermentation of hemicellulose hydrolysates. In: Wall J, Harwood CS, Demain A (eds) Bioenergy. American Society for Microbiology, Washington, pp 37–47

Jeffries TW, Grigoriev IV, Grimwood J et al (2007) Genome sequence of the lignocellulose-bioconverting and xylose-fermenting yeast *Pichia stipitis*. Nat Biotechnol 25(3). doi:10.1038/nbt1290

Jeffries TW, Kurtzman CP (1994) Strain selection, taxonomy, and genetics of xylose-fermenting yeasts. Enzyme Microb Technol 16:922–932

Jeffries TW, van Vleet JRH (2009) *Pichia stipitis* genomics, transcriptomics, and gene clusters. FEMS Yeast Res 9(6):793–807

Kahar P, Tanaka S (2014) A xylose-fermenting yeast hybridized by intergeneric fusion between *Saccharomyces cerevisiae* and *Candida intermediai* mutants for ethanol production. Sustain Chem Process 2:17–28

Karczewska H (1959) Some observations on pentose utilization by *Candida tropicalis*. Compt Rend Lab Carlsberg 31(17):251–258

Koppram R, Albers E, Olsson L (2012) Evolutionary engineering strategies to enhance tolerance of xylose utilizing recombinant yeast to inhibitors derived from spruce biomass. Biotechnol Biofuels 5:32–43

Kordowska-Wiater M (2015) Production of arabitol by yeasts: current status and future prospects. J Appl Microbiol 119:303–314

Krahulec S, Kratzer R, Longus K, Nidetzky B (2012) Comparison of *Scheffersomyces stipitis* strains CBS 5773 and VBS 6054 with regard to their xylose metabolism: implications for xylose fermentation. Microbiol Open 1(1):64–70

Kurtzman CP, Fell JW, Boekhout T (2011) The yeasts, a taxonomic study. Elsevier, Amsterdam

Laluce C, Schenberg ACG, Gallardo JCM, Coradello LFC, Pombeiro-Sponchiado SR (2012) Advances and developments in strategies to improve strains of *Saccharomyces cerevisiae* and processes to obtain the lignocellulosic ethanol—a review. Appl Biochem Biotechnol 166:1908–1926

Lertwattanasakul N, Rodrussamee N, Suprayogi Limtong S, Thanonkeo P, Kosaka T, Yamada M (2011) Utilization capability of sucrose, raffinose and inulin and its less-sensitiveness to glucose repression in thermotolerant yeast *Kluyveromyces marxianus* DMKU3–1042. AMB Express 1:20

Limtong S, Sringiew C, Yongmanitchai W (2007) Production of fuel ethanol at high temperature from sugar cane juice by a newly isolated *Kluyveromyces marxianus*. Bioresour Technol 98:3367–3374

Limtong S, Nitiyon S, Kaewwichian R, Jindamorakot S, Am-In S, Yongmanitchai W (2012) *Wickerhamomyces xylosica* sp. nov. and *Candida phayaonensis* sp. nov., two xylose-assimilating yeast species from soil. Int J Syst Evol Microbiol 62:2786–2792

Lobo F, Gonçalves DL, Alves SL Jr et al (2014) Draft genome sequence of the D-xylose-fermenting yeast *Spathaspora arborariae* UFMG-HM19.1A^T. Genome Announc 2(1):e01163-13. doi:10.1128/genomeA.01163-13

Lorliam W, Akaracharanya A, Suzuki M, Ohkuma M, Tanasupawat S (2013) Diversity and fermentation products of xylose-utilizing yeasts isolated from buffalo feces in Thailand. Microb Environ 28(3):354–360

Margeot A, Hahn-Hagerdal B, Edlund M, Slade R, Monot F (2009) New improvements for lignocellulosic ethanol. Curr Opin Biotechnol 20:372–380

Martini C (2014) Isolamento, identificação e caracterização de linhagem de levedura quanto ao crescimento e fermentação utilizando meios sintéticos com pentoses e hidrolisado de bagaço de cana-de-açúcar. Thesis, Universidade Estadual Paulista Julio de Mesquita Filho

Martini C, Tauk-Tornisielo SM, Codato CB, Bastos RG, Ceccato-Antonini SR (2016) A strain of *Meyerozyma guilliermondii* isolated from sugarcane juice is able to grow and ferment pentoses in synthetic and bagasse hydrolysate media. World J Microbiol Biotechnol 32:80. doi:10.1007/s11274-016-2036-1

Martiniano SE, Chandel AK, Soares LCSR, Pagnocca FC, Silva SS (2013) Evaluation of novel xylose-fermenting yeast strains from Brazilian forests for hemicellulosic ethanol production from sugarcane bagasse. 3 Biotech. doi:10.1007/s13205-013-0145-1

Matos ITSR, Cassa-Barbosa LA, Galvão RSM, Nunes-Silva CG, Astolfi-Filho S (2014) Isolation, taxonomic identification and investigation of the biotechnological potential of wild-type *Meyerozyma guilliermondii* associated with Amazonian termites able to ferment D-xylose. Biosci J 30(1):260–266

Meshitsuka G, Isogai A (1996) Chemical structures of cellulose, hemicellulose, and lignin. In: Hon DNS, Shiraishi N (eds) Chemical modification of lignocellulosic materials. Marcel Dekker, New York, pp 11–34

Milessi TSS, Antunes FAF, Chandel AK, Silva SS (2013) Rice bran extract: an expensive nitrogen source for the production of 2G ethanol from sugarcane bagasse hydrolysate. 3 Biotech 3:373–379

Milessi TSS, Antunes FAF, Chandel AK, Silva SS (2015) Hemicellulosic ethanol production by immobilized cells of *Scheffersomyces stipitisi*: effect of cell concentration and stirring. Bioengineered 6(1):26–32

Miller TL, Churchill BW (1986) Substrates for large-scale fermentation. In: Demain AL, Solomon NA (eds) Manual of industrial microbiology and biotechnology. American Society of Microbiology, Washington, pp 130–131

Morais CG, Cadete RM, Uetanabaro APT, Rosa LH, Lachance MA, Rosa CA (2013) D-xylose-fermenting and xylanase-producing yeast species from rotting wood of two Atlantic Rainforest habitats in Brazil. Fungal Genet Biol 60:19–26

Mouro A (2012) Fermentação de xilose e celobiose por leveduras isoladas da biodiversidade brasileira. Dissertation, Universidade Federal de Santa Catarina

Mussato SI, Roberto IC (2004) Avaliação de diferentes tipos de carvão ativo na destoxificação de hidrolisado de palha de arroz para produção de xilitol. Cienc Tecnol Alim 24:94–100

Náhlik J, Palatová M, Gírio F, Roseiro C (2003) Model identification and physicological control of xylitol production using *Debaryomyces hansenii*. Process Biochem 38(12):1695–1705

Nguyen NH, Suh SO, Marshall CJ, Blackwell M (2006) Morphological and ecological similarities: wood-boring beetles associated with novel xylose-fermenting yeasts, *Spathaspora passalidarum* gen. sp. nov. and *Candida jeffriesii* sp. nov. Mycol Res 110:1232–1241

Nigam JN, Ireland RS, Margaritis A, Lachance MA (1985) Isolation and screening of yeasts that ferment D-xylose directly to ethanol. Appl Environ Microbiol 50(6):1486–1489

Nitiyon S, Boonmak C, Am-In S, Jindamorakot S, Kawasaki H, Yongmanitchai W, Limton S (2011) *Candida saraburiensis* sp. nov. and *Candida prachuapensis* sp. nov., xylose-utilizing yeast species isolated in Thailand. Int J Syst Evol Microbiol 61:462–468

Nitiyon S, Keo-Oudone C, Murata M, Lertwattanasakul N, Limtong S, Kosaka T, Yamada M (2016) Efficient conversion of xylose to ethanol by stress-tolerant *Kluyveromyces marxianus* BUNL-21. SpringerPlus 5:185

Novy V, Krahulec S, Longus K, Klimacek M, Nidetzky B (2013) Co-fermentation of hexose and pentose sugars in a spent sulfite liquor matrix with genetically modified *Saccharomyces cerevisiae*. Bioresour Technol 130:439–448

Palmqvist E, Hahn-Hagerdal B (2000) Fermentation of lignocellulosic hydrolyzate I: inhibition and detoxication. Bioresour Technol 74:17–24

Parawira W, Tekere M (2011) Biotechnological strategies to overcome inhibitors in lignocellulose hydrolysates for ethanol production: a review. Crit Rev Biotechnol 31(1):20–31

Peng B, Shen Y, Li X, Chen X, Hou J, Bao X (2012) Improvement of xylose fermentation in respiratory-deficient xylose-fermenting *Saccharomyces cerevisiae*. Metab Eng 14(1):9–18

Prompt AH (2012) Análise da fermentação de glucose e xilose por leveduras Spathaspora isoladas de madeira em decomposição. Dissertation, Universidade Federal de Santa Catarina

Rao RS, Bhadra B, Shivaji S (2008) Isolation and characterization of ethanol-producing yeasts from fruits and tree barks. Lett Appl Microbiol 47:19–24

Ren Y, Chen L, Niu Q, Hui F (2014) Description of *Scheffersomyces henanensis* sp. nov., a new D-xylose-fermenting yeast species isolated from rotten wood. PLoS One 9(3):e92315. doi:10.1371/journal.pone.0092315

Richard P, Verho R, Putkonen M, Londesborough J, Penttila M (2003) Production of ethanol from L-arabinose by *Saccharomyces cerevisiae* containing a fungal L-arabinose pathway. FEMS Yeast Res 3:185–189

Runquist D, Hahn-Hagerdal B, Bettiga M (2009) Increased expression of the oxidative pentose phosphate pathway and gluconeogenesis in anaerobically growing xylose-utilizing *Saccharomyces cerevisiae*. Microb Cell Fact 8:49–61

Sampaio JP (1999) Utilization of low molecular weight aromatic compounds by heterobasidiomycetous yeasts: taxonomic implications. Can J Microbiol 45(6):491–512

Sanchez RG, Karhumaa K, Fonseca C et al (2010) Improved xylose and arabinose utilization by an industrial recombinant *Saccharomyces cerevisiae* strain by using evolutionary engineering. Biotechnol Biofuels 3:13–23

Sanchez S, Bravo V, Castro E, Moya AJ, Camacho F (1997) The influence of pH and aeration rate on the fermentation of D-xylose by *Candida shehatae*. Enzyme Microb Technol 21:355–360

Sanchez S, Bravo V, Castro E, Moya AJ, Camacho F (2002) The fermentation of mixtures of D-glucose and D-xylose by *Candida shehatae, Pichia stipitis* or *Pachysolen tannophilus* to produce ethanol. J Chem Technol Biotechnol 77(6):641–648

Sanchez S, Bravo V, Castro E, Moya AJ, Camacho F (2004) Influence of temperature on the fermentation of d-xylose by *Pachysolen tannophilus* to produce ethanol and xylitol. Process Biochem 39:673–679

Sarks C, Mingjie J, Sato TK, Balan V, Dale BE (2014) Studying the rapid bioconversion of lignocellulosic sugars into ethanol using high cell density fermentations with cell recycle. Biotechnol Biofuels 7:73–84

Sasaki M, Inui M, Yukawa H (2012) Microorganisms for xylitol production: focus on strain improvement. In: Silva SS, Chandel AK (eds) D-xylitol: fermentative production, application and commercialization. Springer, Heidelberger, pp 109–132

Schneider H, Wang PY, Yk C, Maleszka R (1981) Conversion of D-xylose into ethanol by the yeast *Pachysolen tannophilus*. Biotechnol Lett 3(2):89–92

Skoog K, Hahn-Hagerdal B (1990) Effect of oxygenation on xylose fermentation by *Pichia stipitis*. Appl Environ Microbiol 56(11):3389–3394

Signori L, Passolunghi S, Ruohonen L, Porro D, Branduardi P (2014) Effect of oxygenation and temperature on glucose-xylose fermentation in *Kluyveromyces marxianus* CBS712 strain. Microb Cell Fact 13:51

Silva DV, Cândido EJ, Arruda PV, Silva SS, Felipe MGA (2014) New culture medium for bioconversion of C5 fraction from sugarcane bagasse using rice bran extract. Braz J Microbiol 45(4):1469–1475

Silva JPA, Mussatto SI, Roberto IC (2010) The influence of initial xylose concentration, agitation, and aeration on ethanol production by *Pichia stipitis* from rice straw hemicellulosic hydrolysate. Appl Microbiol Biotechnol 162:1306–1315

Silva JPA, Mussatto SI, Roberto IC, Teixeira JA (2011) Ethanol production from xylose by *Pichia stipitis* NRRL Y-7124 in a stirred tank bioreactor. Braz J Chem Eng 28(1):151–156

Silva JPA, Mussatto SI, Roberto IC, Teixeira JA (2012) Fermentation medium and oxygen transfer conditions that maximize the xylose conversion to ethanol by *Pichia stipitis*. Renew Energy 37:259–265

Slininger PJ, Bothast RJ, Ladisch MR, Okos MR (1990) Optimum pH and temperature conditions for xylose fermentation by *Pichia stipitis*. Biotechnol Bioeng 35(7):727–731

Smith J, van Rensburg E, Gorgens JF (2014) Simultaneously improving xylose fermentation and tolerance to lignocellulosic inhibitors through evolutionary engineering of recombinant *Saccharomyces cerevisiae* harbouring xylose isomerase. BMC Biotechnol 14:41–57

Soccol CR, Vandenberghe LPS, Medeiros ABP, Karp SG, Buckeridge MS, Ramos LP, Pitarelo AP, Ferreira-Leitão V, Gottschalk LMF, Ferrara MA, Bon EPS, Moraes LMP, Araújo JA, Torres FAG (2010) Bioethanol from lignocelluloses: status and perspectives in Brazil. Bioresour Technol 101:4820–4825

Souto-Maior AM, Runquist D, Hahn-Hagerdal B (2009) Crabtree-negative characteristics of recombinant xylose-utilizing *Saccharomyces cerevisiae*. J Biotechnol 143:119–123

Sreenath HK, Jeffries TW (2000) Production of ethanol from wood hydrolyzate by yeast. Bioresour Technol 72:253–260

Su YK, Willis LB, Jeffries TW (2015) Effects of aeration on growth, ethanol and polyol accumulation by *Spathaspora passalidarum* NRRL Y-27907 and *Scheffersomyces stipitis* NRRL Y-7124. Biotechnol Bioeng 112(3): 457–469. doi:10:1002/bit.25445

Suh SO, Marshall CJ, McHugh JV, Blackwell M (2003) Wood ingestion by passalid beetles in the presence of xylose-fermenting gut yeasts. Mol Ecol 12:3137–3145

Suh SO, White MM, Nguyen NH, Blackwell M (2004) The status and characterization of *Enteroramus dimorphus*: a xylose-fermenting yeast attached to the gut of beetles. Mycologia 96(4):756–760

Talebnia F, Taherzadeh MJ (2006) In situ detoxification and continuous cultivation of dilute-acid hydrolyzate to ethanol by encapsulated *Saccharomyces cerevisiae*. J Biotechnol 125(3):377–384

Tanimura A, Nakamura T, Watanabe I, Ogawa J, Shima J (2012) Isolation of a novel strain of *Candida shehatae* for ethanol production at elevated temperature. SpringerPlus 1:27–33

Tengborg C, Galbe M, Zacchi G (2001) Influence of enzyme loading and physical parameters on the enzymatic hydrolysis of steam-pretreated softwood. Biotechnol Prog 17(1):110–117

Toivola A, Yarrow D, van den Bosch E, van Dijken JP, Scheffer WA (1984) Alcoholic fermentation of D-xylose by yeasts. Appl Environ Microbiol 47:1221–1223

Tomme P, Warren RAJ, Gilkes NR (1995) Cellulose hydrolysis by bacteria and fungi. Adv Microb Physiol 37:1–81

Urbina H, Schuster J, Blackwell M (2013) The gut of Guatemalan passalid beetles: a habitat colonized by cellobiose- and xylose-fermenting yeasts. Fungal Ecol 6:339–355

van der Walt JP, Ferreira NP, Steyn RL (1987) *Candida lyxosophila* sp. nov., a new D-xylose fermenting yeast from Southern Africa. Antonie van Leeuwenhoek 53:93–97

Vandeska E, Amartey S, Kuzmanova S, Jeffries TW (1995) Effects of environmental conditions on production of xylitol by *Candida boidinii*. World J Microbiol Biotechnol 11:213–218

Varize CS (2013) Isolamento e seleção de leveduras para fermentação de xilose. Dissertation, Universidade de São Paulo – ESALQ

Wallace-Salinas V, Gorwa-Grauslund MF (2013) Adaptive evolution of an industrial strain of *Saccharomyces cerevisiae* for combined tolerance to inhibitors and temperature. Biotechnol Biofuels 6:151–159

Wei N, Quaterman J, Kim SR, Cate JHD, Jin YS (2013) Enhanced biofuel production through coupled acetic acid and xylose consumption by engineered yeast. Nat Commun 4:2580. doi:10.1038/ncomms3580

Wisselink HW, Toirkens MJ, Berriel MRF, Winkler AA, van Dijken JP, Pronk JT, van Maris AJA (2007) Engineering of *Saccharomyces cerevisiae* for efficient anaerobic alcoholic fermentation of L-arabinose. Appl Environ Microbiol 73(15):4881–4891

Wisselink HW, Toirkens MJ, Wu Q, Pronk JT, van Maris AJA (2009) Novel evolutionary engineering approach for accelerated utilization of glucose, xylose, and arabinose mixtures by engineered *Saccharomyces cerevisiae* strains. Appl Environ Microbiol 75(4):907–914

Yang B, Wyman CE (2008) Pretreatment: the key to unlocking low-cost cellulosic ethanol. Biofuels Bioprod Biorefin 2(1):26–40

Zhang W, Geng A (2012) Improved ethanol production by a xylose-fermenting recombinant yeast strains constructed through a modified genome shuffling method. Biotechnol Biofuels 5:46–56

Chapter 9
New Developments in Sugarcane Genetics and Genomics

Melina Cristina Mancini, Cláudio Benício Cardoso-Silva, Estela Araujo Costa, Thiago Gibbin Marconi, Antonio Augusto Franco Garcia, and Anete Pereira De Souza

Abstract Modern sugarcane cultivars (*Saccharum* spp.) are derived from an interspecific hybridization between *Saccharum officinarum* and *Saccharum spontaneum* and pose a significant challenge for both genotyping and data analysis. Due to their large (estimated at approximately 10 Gb) and complex genome, which can include variable ploidy levels and aneuploidy, studies involving molecular markers for genetic and quantitative trait locus (QTL) mapping are extremely laborious. Several advances in the genetics and genomics of sugarcane have recently become possible with the emergence of new sequencing technologies, the use of several types of marker systems, and the use of genotyping data analysis software. Molecular markers and comparative genomics are powerful resources that allow us to explore allelic variation and to thus understand the genome organization of sugarcane. This chapter provides an overview of what is known about the genetic structure and the genomics of sugarcane as well as the main genomics strategies developed for sugarcane. Among the strategies discussed are the use of single-nucleotide polymorphisms (SNPs) and bacterial artificial chromosome (BAC) libraries and the analysis of the syntenic relationships with related species (maize, sorghum, and rice).

Keywords Bacterial artificial chromosome • Genetic mapping • Molecular markers • Transcriptome

---

M.C. Mancini • C.B. Cardoso-Silva • E.A. Costa • T.G. Marconi
Centro de Biologia Molecular e Engenharia Genetica (CBMEG), Universidade Estadual de Campinas (UNICAMP), Campinas, SP, Brazil

A.A.F. Garcia
Departamento de Genetica, Escola Superior de Agricultura Luiz de Queiroz, Universidade de São Paulo, Piracicaba, SP, Brazil

A.P. De Souza (✉)
Centro de Biologia Molecular e Engenharia Genetica (CBMEG), Universidade Estadual de Campinas (UNICAMP), Campinas, SP, Brazil

Departamento de Biologia Vegetal, Instituto de Biologia, Universidade Estadual de Campinas (UNICAMP), Campinas, SP, Brazil
e-mail: anete@unicamp.br

M.S. Buckeridge, A.P. De Souza (eds.), *Advances of Basic Science for Second Generation Bioethanol from Sugarcane*, DOI 10.1007/978-3-319-49826-3_9

9.1 Introduction

Currently, the increasing worldwide use of biofuels is one of the main factors responsible for financial and scientific investments in research. Since the early 2000s, researchers from UNICAMP (Universidade Estadual de Campinas, Brazil), ESALQ-USP (Universidade de São Paulo, Escola Superior de Agricultura Luiz de Queiroz, Piracicaba/SP), IAC (Instituto Agronômico de Campinas, Campinas/SP), and UFSCar (Universidade Federal de São Carlos, Araras/SP) have focused extensively on several related areas to advance knowledge of the genetic structure and genomics of sugarcane. The research of these groups is primarily dedicated to microsatellites or SSRs (Simple Sequence Repeats; Tautz 1989) and SNPs (Single-Nucleotide Polymorphisms), linkage maps and quantitative trait loci (QTL) mapping, the construction and evaluation of bacterial artificial chromosome (BAC) libraries, and the transcriptome analyses of several tissues. The first sugarcane research published by these groups was an analysis of genetic similarity and coefficient of parentage (Lima et al. 2002), where the authors used the polymorphisms revealed by AFLP (Amplified Fragment Length Polymorphism; Vos et al. 1995) to identify the genetic similarities among 79 sugarcane genotypes.

The development of molecular markers, particularly SSRs, is an important tool used by these groups. Several scientific reports were published, with more than 700 novel SSRs (Pinto et al. 2004, 2006; Oliveira et al. 2009; Marconi et al. 2011), which were identified using the SUCEST database (sugarcane ESTs; Vettore et al. 2001, 2003). In 2013, the development of approximately 1000 SNPs was used to study dosage effects in sugarcane using a novel approach for genetics and molecular studies (Garcia et al. 2013). This study represents an important advancement in the understanding of sugarcane genetics.

Several new statistical methods related to linkage maps and QTL mapping were developed and resulted in significant contributions to the scientific community (Garcia et al. 2006; Oliveira et al. 2007; Pastina et al. 2012; Gazaffi et al. 2014; Margarido et al. 2015). These researchers cataloged more than 5000 novel unigenes in sugarcane leaves and identified a large number of molecular markers including 5106 SSRs and more than 700,000 SNPs (Cardoso-Silva et al. 2014).

Also, two bacterial artificial chromosome (BAC) libraries for the SP80-3280 and IACSP96-3046 varieties were constructed, with 2.4 and 1.8 times the genome coverage, respectively, representing the biggest BAC libraries from sugarcane to date. These BAC libraries make it possible to determine the architecture of different loci with variable ploidy and also allow for the study of regions associated with interesting traits.

9.2 General Aspects of Sugarcane and the Use of Molecular Markers

Modern sugarcane varieties result from interspecific crosses between *Saccharum officinarum* (basic chromosome number: $x = 10$; $2n = 8x = 80$) and *Saccharum spontaneum* (basic chromosome number: $x = 8$; $5x \leq 2n \leq 16x$; $40 \leq 2n \leq 128$). This process enabled asymmetric chromosome transmission, resulting in varieties with a different chromosome number in somatic cells ($2n$), generally 100–130 chromosomes (D'Hont et al. 1998; Irvine 1999; Grivet and Arruda 2001). The aneuploidy condition and high ploidy level make sugarcane a crop plant with high genetic complexity.

The use of molecular biology tools, particularly molecular markers, has increased our understanding of the relevance of genetic polymorphisms and has been further enabled by studies on DNA markers that are genetically linked to traits of interest. The main markers used in molecular genetics studies are RFLP (Restriction Fragment Length Polymorphism; Botstein et al. 1980), AFLP, SSR, and, more recently, SNP. These markers are widely used in the construction of sugarcane genetic maps (Al-Janabi et al. 1993; Silva et al. 1993, 1995; D'Hont et al. 1994; Grivet et al. 1996; Mudge et al. 1996; Ming et al. 1998, 2002a; Guimarães et al. 1997; Guimarães 1999; Hoarau et al. 2001; Aitken et al. 2005, 2007; Raboin et al. 2006; Garcia et al. 2006; Oliveira et al. 2007; Palhares et al. 2012) and for the identification of genomic regions associated with QTLs (Sills et al. 1995; Daugrois et al. 1996; Ming et al. 2001, 2002b, 2002c; Hoarau et al. 2002; Jordan et al. 2004; Silva and Bressiani 2005; McIntyre et al. 2005a, 2005b, 2006; Reffay et al. 2005; Aitken et al. 2006, 2008; Raboin et al. 2006; Wei et al. 2006; Al-Janabi et al. 2007; Piperidis et al. 2008; Pinto et al. 2010; Pastina et al. 2012).

9.3 Identification of Molecular Markers in Sugarcane

Databases are an excellent resource for the development of molecular markers and are available for several crop species. In the special case of sugarcane, a consortium of Brazilian researchers, called the SUCEST project, generated approximately 238,000 expressed sequences by sequencing cDNA libraries from several tissues (Vettore et al. 2003). These sequences were used to identify and develop microsatellite markers (Pinto et al. 2004, 2006; Oliveira et al. 2009; Marconi et al. 2011). Furthermore, several differentially expressed contigs identified by SUCEST were used to develop SNP markers (Garcia et al. 2013). All these molecular markers were used in the construction of genetic linkage maps and for QTL mapping (Pinto et al. 2010; Pastina et al. 2012; Santos et al. 2014).

Although the functional markers are more informative due to their direct association with expressed regions of the genome, they are less abundant and polymorphic than genomic markers. To construct a genetic map, it is essential to identify molecular markers scattered throughout the genome. Therefore, genotyping using genomic markers is necessary to construct a saturated genetic map.

With the availability of genomic sequences from BAC libraries containing large fragments of genomic DNA, the development of many molecular markers will be possible. The BAC sequencing provides a manner of splitting the sugarcane genome into several parts which makes genome assembly easier. Furthermore, these sequences contain many repetitive regions and can be used to search for molecular markers, such as microsatellites.

Due to their importance, genomic sequences originating from BACs sequenced by Setta and collaborators (2014) were mined for microsatellite regions. In these data 4342 microsatellites were identified within 280 BAC sequences spanning 32 Mbp. When we consider microsatellite types, the most abundant are di- and trinucleotide, which represent 57% of the total.

The distribution of microsatellites was mapped using genomic sequences from grasses, such as sugarcane, maize, sorghum, and rice. When sugarcane was compared with sorghum and maize, we noticed that the frequency of microsatellite motif distribution was similar among these species (Cardoso-Silva 2015). Although the sugarcane genome is not complete, the results are still in accordance with the phylogenetic relationships among these grasses. When we consider microsatellites that occur in coding regions, only 686 motifs were identified; of those, 161 were exonic and 525 were intronic (Cardoso-Silva 2015). The trinucleotide microsatellites were more abundant in exons. In fact, previous studies have shown that trinucleotide repeats are primarily exonic for all eukaryotes studied (Toth et al. 2000; Morgante et al. 2002) and constitute approximately 93% of the SSRs in monocots (Sonah et al. 2011). It has been hypothesized that the prevalence of trinucleotide microsatellites in exons is because this motif may offer a lower risk of frameshift mutations by maintaining the amino acid sequence of a protein if point mutations occur (Metzgar et al. 2000).

The development of molecular markers from expressed sequence tags (ESTs) and genomic sequences is fundamentally important for potential applications in breeding programs. The genomic markers are characterized by their high polymorphism level, whereas the expressed sequences are required due to their direct association with functional regions in the genome. Additionally, the usage of molecular markers directly associated with candidate genes can accelerate the identification of QTLs and their application in breeding programs.

9.4 Genetic Mapping in Sugarcane

In some fruit plants (e.g., passion fruit, cocoa, and apple) and forest trees (e.g., rubber tree, populus, and eucalyptus), obtaining inbred lines is impractical due to the amount of time necessary to achieve homozygotes or because of inbreeding depression. In this case, a genetic map is constructed using an F_1 segregating population, based on a full-sib progeny originating from a cross between two non-inbred parents. To construct genetic maps in this situation, a previously commonly used alternative is the so-called double pseudo-test cross, resulting in the construction of two individual maps through the identification of polymorphisms in each parent (Grattapaglia and Sederoff 1994; Shepherd et al. 2003; Porceddu et al. 2002; Carlier et al. 2004).

Based on this approach and using only single dosage markers, linkage maps for sugarcane, *S. officinarum* ("LA Purple") and *S. robustum* ("Mol 5829") were constructed using RAPD (Random Amplified Polymorphic DNA; Williams et al.

1990), RFLP, and AFLP markers (Guimarães 1999). However, the information contained in these individual maps can be integrated into a single map using heterozygous markers (3:1 segregation) in both parents, which are used to establish a link between the markers that segregate individually for each parent (Barreneche et al. 1998; Wu et al. 2000; Garcia et al. 2006; Oliveira et al. 2007; Palhares et al. 2012). The construction of an integrated genetic map using markers with different types of segregation has a number of advantages because it allows for the saturation of the linkage map and extends the characterization of polymorphic variation to the entire genome. Specifically, for polyploid species, codominant markers can be useful for clustering co-segregated groups into their respective homology groups (Silva et al. 1993). For sugarcane, integrated maps were generated by Garcia et al. (2006), Oliveira et al. (2007), and Palhares et al. (2012) using the approach proposed by Wu et al. (2002) and the R package described by Margarido et al. (2007). These maps were based on a full-sib population derived from the cross between two pre-commercial varieties, SP80-180 (female parent) and SP80-4966 (male parent).

More recently, SNPs have been used as markers, resulting in novel perspectives on disease diagnosis, varietal characterization, and marker-assisted selection, as well as in the construction of high-resolution genetic maps and association mapping (Remington et al. 2001; Rafalski 2002; Batley et al. 2003; Wilson et al. 2004; Yu and Buckler 2006; Agrama et al. 2007; Salvi et al. 2007). SNPs are variations in the DNA sequence that occur when a single nucleotide is altered. These polymorphisms, along with deletions and insertions, are responsible for the majority of variation found in organisms (Cho et al. 1999; Rafalski and Tingey 2008) and are widely distributed throughout the genome. However, SNPs are more abundant in non-transcribed regions and regions flanking SSRs (Bundock and Henry 2004; Mogg et al. 2002). They are abundant in many species, including maize (Tenaillon et al. 2001), barley (Kanazin et al. 2002), and rice (Yu et al. 2002). With improvements in and increased affordability of next-generation sequencing technologies (and expressed sequence databases), the identification and use of SNPs is increasing. For sugarcane, the great advantage of SNPs over other markers types is the possibility of determining the copy number of each allele (marker dosage) behaving as a codominant marker.

Molecular markers, except for SNPs, with a codominant inheritance in diploid organisms, behave like dominant markers in sugarcane. Identifying potential polymorphisms for most markers is of limited usage in sugarcane because of its high ploidy level. Consequently, the genetic maps currently available for sugarcane without the use of SNPs did not achieve adequate coverage. These maps are estimated to represent only 33–50% of the genome. By using SNPs as molecular markers, higher resolution and more saturated maps can be produced due to the abundance of SNPs and the high degree of automation involved in their analysis. Additionally, they have a co dominant nature in polyploid organisms (Nasu et al. 2002; Garcia et al. 2013), as already shown in barley (Rostoks et al. 2005) and sunflower (Lai et al. 2005).

9.5 SNP Genotyping Using Mass Spectrometry (Sequenom MassARRAY®)

In recent years, matrix-assisted laser desorption/ionization time-of-flight mass spectrometry (MALDI-TOF MS), the technology used by Sequenom iPLEX MassARRAY® (Sequenom 2007), has emerged as a powerful method for genotyping SNPs. Its main attributes are precision, speed in generating data, the possibility of performing multiple reactions, and the resulting data structure. Additionally, mass spectrometry is a powerful and versatile analytical method that provides valuable information about the composition and structure of molecules and can distinguish a number of specific analytes in mixtures. It can accurately detect and quantify allele frequencies as low as a few percent (Oberacher 2008). Due to an increase in the demand for SNP genotyping and a reduction in the cost, many methods for genotyping SNPs with MALDI-TOF MS have been implemented and, with a high degree of automation, have been applied to large-scale studies (Gut 2004). For polyploid organisms, the MassARRAY® System has proven to be very useful because it can measure different allele frequencies within the polymorphic site.

Recent advances in next-generation DNA sequencing technology have led to a revolution in the way populations are genotyped. One promising approach called genotyping-by-sequencing (GBS) is based on high-throughput sequencing of genomic subsets targeted by restriction enzymes (Elshire et al. 2011). This approach has been demonstrated to be robust and capable of producing many thousands of molecular markers, such as SNPs, and is suitable for breeding and trait mapping in diverse organisms. GBS has been useful in the characterization of crop plants with complex genomes, such as wheat (Poland et al. 2012) and oat (Huang et al. 2014). In sugarcane, in addition to SNP identification by mass spectrometry, the GBS strategy has already been used to produce SNP markers. However, the success of this strategy is still dependent on the development of downstream analysis tools for complex polyploids such as sugarcane.

In polyploid organisms, the frequency of SNPs in a certain locus is determined by the number of chromosomes that carry that specific locus, by the number of different alleles, and by the frequency of each allele. However, at least for sugarcane, little is known about the degree of similarity between homologous chromosomes within homology groups or about the complexity and allelic diversity at a single locus of a single gene. As a result, any method used to detect SNPs in a certain sugarcane locus must be able to determine the frequency of each base in different genotypes, rather than just detecting the presence or absence of SNPs; this is exactly what occurs in MALDI-TOF MS. Such detection systems are more complex and expensive, but the information generated by them is essential to better understand the genetic and genomic structure in polyploids. In sugarcane, a polymorphic SNP between two genotypes is defined when an allele is present in one individual and absent in another or when the genotypes have different alleles or a different allele copy number (dose). Because the allele copy number is proportional to the height of each spectrometry peak (allele abundance), it is possible to estimate the number of copies. Data classification for polyploidy is a complex issue which has led to the development of a new statistical method (SNP calling) and new software (SuperMASSA) for the analysis and classification of SNP data generated by mass spectrometry (Serang et al. 2012).

9.6 Analysis of the Sugarcane Genome Using SNPs

Garcia et al. (2013) reported the development and evaluation of SNPs for sugarcane. As a result, a new methodology for quantitative analysis allowing the allelic dosage estimation was applied for the first time. The development of this methodology led to the discovery of important information about the complex architecture of the sugarcane genome. Several SNPs (1041) were genotyped in an association-mapping panel with relevant sugarcane genotypes and 271 SNPs in the progeny of a biparental population derived from a cross between the Brazilian varieties IACSP95-3018 (female parent) and IACSP93-3046 (male parent). All the SNPs were genotyped using a method based on MALDI-TOF analysis, which was performed on a mass spectrometer platform from Sequenom MassARRAY® using iPLEX GOLD chemistry.

The SNP assay is based on a single-base extension of the primer, which is locus specific, with a mass-modified nucleotide. The mass intensities for each individual in the population can be represented in a scatter plot (Fig. 9.1). Assuming equal efficiency in the reaction, the mass intensities can represent the abundance of each allele in the specific locus. In Fig. 9.1, it is possible to observe the formation of clusters with a different number of points (individuals). If the ploidy level is known, the number of clusters can be used to estimate the dosage. In diploids, for example, the scatter plot should contain three clusters, for both homozygotes and heterozygotes (not considering copy number variation), with expected angles of 0°, 45°, and 90° (indicating intensity ratios of (2,0), (1,1), and (0,2)). For autotetraploids, there should be five possible results (from 0 to 4 copies of a given allele) and so on, depending on the ploidy level.

In general, for sugarcane, the situation is more complicated because the number of clusters will depend on the ploidy level which is unknown. So, it was necessary to estimate ploidy without reliable cytological studies. The estimation of sugarcane ploidy was achieved using the method of Serang et al. (2012) that simultaneously considers the combination of parental information, the number of clusters, the intensities associated with different alleles, and the expected frequencies of individuals in each cluster. This approach was used to estimate the ploidy level in the SNPs genotyped in the biparental cross (IACSP95-3018 and IACSP93-3046). The results indicated that their ploidy level was within the 6–12 range. It is remarkable that only a few of the SNPs were present as single-dose (SD) alleles (21 out of the 241 SNPs). Because most sugarcane genetics studies have only considered simplex markers, the actual sample of the genome already explored is rather small (Garcia et al. 2013). The observation of a small number of SD markers is quite different from previously published results in sugarcane (Hoarau et al. 2001; Aitken et al. 2005; George and Aitken 2010, among others).

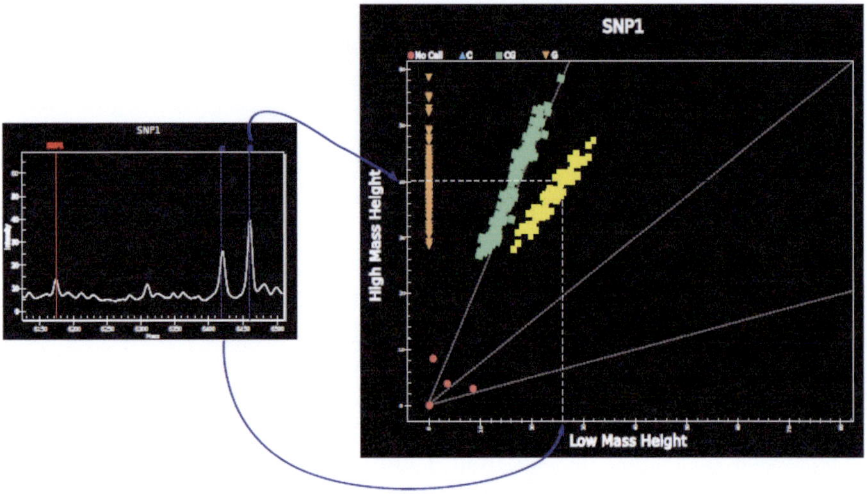

Fig. 9.1 Example of SNP genotyping using Sequenom MassARRAY®. *Left*: mass intensity of each allele of a given SNP. *Right*: each dot is a different individual allocated in the scatter plot depending on mass intensity for the two alleles (with low and high mass)

9.7 Transcriptome Analysis

Transcriptome studies in sugarcane have primarily relied on public EST datasets. The first dataset was reported by researchers in South Africa (Carson and Botha 2000, 2002), but the largest EST collection is the Brazilian SUCEST project, which has 238,000 ESTs (Vettore et al. 2001, 2003). There are also EST libraries in Australia (Casu et al. 2003, 2004; Bower et al. 2005) and the United States (Ma et al. 2004), with approximately 10,000 ESTs each. The availability of large collections of expressed sequence tags (ESTs) provides information that can possibly be used for functional molecular marker development.

Recent technological developments in NGS have the potential to significantly increase our understanding of the organization of complex genomes. The use of RNA-Seq technology in large complex genomes has the major advantage of substantially reducing the effective genome size. For example, the 2.3 Gb haploid maize genome is reduced to a 97 Mb transcriptome, assuming all genes are expressed (Hansey et al. 2012).

RNA-Seq is a powerful method used to map and quantify transcriptomes, which was developed to analyze global gene expression in different tissues. This technique has also been used as an efficient and cost-effective method of systematically identifying SNPs in transcribed regions for different species (Cloonan et al. 2008; Morin et al. 2008; Chepelev et al. 2009; Cirulli et al. 2010). Furthermore, unprecedented opportunities arose for high-throughput functional genomics research, including gene expression profiling, genome annotation, and small ncRNA discovery (Bentley

2006; Morozova and Marra 2008). In this way, NGS technologies could have significant implications for crop genetics and breeding. This technology has been applied as a tool for RNA sequencing in many species, such as *Arabidopsis thaliana* (Lister et al. 2008), *Brassica* spp. (Trick et al. 2009), rice (Lu et al. 2010), maize (Hansey et al. 2012), rubber tree (Mantello et al. 2014), and *Panicum maximum* (Toledo-Silva et al. 2013). Through the use of NGS technologies such as RNA-Seq, transcriptome analysis has proven to be a valuable first step in the study of genetic traits and has permitted researchers to obtain sequence information and expression levels for genes involved in important metabolic pathways and to also discover molecular markers in model and non-model organisms (Van Belleghem et al. 2012).

The first whole transcriptome from commercial sugarcane varieties was recently produced using NGS technology (Cardoso-Silva et al. 2014). Gene annotation based on Gene Ontology enrichment analysis allowed the identification of several transcripts with functions that were distinctly represented among the different sugarcane genotypes. For example, the identification of enriched terms involved in stimulus-response pathways allowed the comparison of genotypes enriched with resistance-related genes with the genotypes that presented resistance traits. All of the genotypes were enriched for transcripts involved in the network of sucrose synthesis, accumulation, storage, and retention, which was consistent with the high level of sucrose production in these sugarcane genotypes.

9.8 Conclusions and Perspectives

Construction of the sugarcane genetic map and QTL mapping based on marker systems is still incomplete and challenging due to the high ploidy level that occurs in the sugarcane genome. Alternatively, BAC approaches are being used to understand the sugarcane genome organization by sequencing a long DNA fragment of interest. This approach allows for the investigation of tagged genes as well as their allelic variation originating from causes such as point mutations, insertions/deletions, inversions, and transposons. Furthermore, the QTL regions can be sequenced, and their gene composition can be determined; then, individual alleles can be sampled, and genetic implications can be inferred from sequence variations. Currently, this approach is being used by our group to investigate QTLs of interest in sugarcane varieties, specifically by BAC sequencing from IACSP93-3046 and SP80-3280 libraries

Acknowledgments This research was supported by Conselho Nacional de Desenvolvimento Científico e Tecnológico (CNPq, Grant *145478/2012-2*) and Fundação de Amparo a Pesquisa do Estado de São Paulo (FAPESP, Grant 2005/55258-6, 2008/52197-4, 2010/50031-1, 2010/06715-3, 2010/50549-0). It was also part of the research of the Instituto Nacional de Ciência e Tecnologia do Bioetanol (granted by CNPq, 574002/2008-1, and FAPESP, 2008/57908-6), Coordenação de Aperfeiçoamento de Pessoal de Nível Superior (CAPES, Computacional Biology Program) and the CeProBio Project (granted by CNPq). A.A.F. Garcia and A.P. Souza are recipients of research fellowships from CNPq.

References

Agrama HA, Eizenga GC, Yan W (2007) Association mapping of yield and its components in rice cultivars. Mol Breed 19(4):341–356

Aitken KS, Jackson PA, Mcintyre CL (2005) A combination of AFLP and SSR markers provides extensive map coverage and identification of homo(eo)logous linkage groups in a sugarcane cultivar. Theor Appl Genet 110(5):789–801

Aitken KS, Jackson PA, Mcintyre CL (2006) Quantitative trait loci identified for sugar related traits in a sugarcane (*Saccharum* spp.) cultivar x *Saccharum officinarum* population. Theor Appl Genet 112(7):1306–1317

Aitken KS, Jackson PA, Mcintyre CL (2007) Construction of a genetic linkage map for *Saccharum officinarum* incorporating both simplex and duplex markers to increase genome coverage. Genome 50(8):742–756

Aitken KS, Hermann S, Karno K, Bonnett GD, McIntyre LC, Jackson PA (2008) Genetic control of yield related stalk traits in sugarcane. Theor Appl Genet 117(7):1191–1203

Al-Janabi SM, Honeycutt RJ, McClelland M, Sobral BW (1993) A genetic Linkage Map of *Saccharum spontaneum* L, "SES 208". Genetics 134(4):1249–1260

Al-Janabi SM, Parmessur Y, Kross H, Dhayan S, Saumtally S, Ramdoyal K, Autrey LJC, Dookun-Saumtally A (2007) Identification of a major quantitative trait locus (QTL) for yellow spot (*Mycovellosiella koepkei*) disease resistance in sugarcane. Mol Breed 19(1):1–14

Barreneche T, Bodenes C, Lexer C, Trontin JF, Fluch S, Streiff R, Plomion C, Roussel G, Steinkellner H, Burg K, Favre JM, Glössl J, Kremer A (1998) A genetic linkage map of Quercus robur L. (pedunculate oak) based on RAPD, SCAR, microsatellite, minisatellite, isozyme and 5S rDNA markers. Theor Appl Genet 97(7):1090–1103

Batley J, Barker G, O'Sullivan H, Edwards KJ, Edwards D (2003) Mining for single nucleotide polymorphisms and insertions/deletions in maize expressed sequence tag data. Plant Physiol 132(1):84–91

Bentley DR (2006) Whole-genome re-sequencing. Curr Opin Genet Dev 16:545–552

Botstein D, White RL, Skolnick M, Davis RW (1980) Construction of a genetic linkage map in man using restriction fragment length polymorphisms. Am J Hum Genet 32:314–331

Bower NI, Casu RE, Maclean DJ, Reverter A, Chapman SC, Manners JM (2005) Transcriptional response of sugarcane roots to methyl jasmonate. Plant Sci 168:761–772

Bundock PC, Henry RJ (2004) Single nucleotide polymorphism, haplotype diversity and recombination in the Isa gene of barley. Theor Appl Genet 109(3):543–551

Cardoso-Silva CB (2015) Análise do transcriptoma e de sequências genômicas de variedades comerciais de cana-de-açúcar. Campinas: UNICAMP. Thesis (Doutorado em Genética e Biologia Molecular), Instituto de Biologia, Universidade Estadual de Campinas

Cardoso-Silva CB, Costa EA, Mancini MC, Balsalobre TWA, Canesin LEC, Pinto LR, Carneiro MS, Garcia AAF, Souza AP, Vicentini R (2014) *De Novo* assembly and transcriptome analysis of contrasting sugarcane varieties. PLoS One 9(2):e88462. doi:10.1371/journal.pone.0088462

Carlier JD, Reis A, Duval MF, Coppens D'Eeckenbrugge G, Leitão JM (2004) Genetic maps of RAPD, AFLP and ISSR markers in Ananas bracteatus and A. comosus using the pseudo-testcross strategy. Plant Breed 123(2):186–192

Carson DL, Botha FC (2000) Preliminary analysis of expressed sequence tags for sugarcane. Crop Sci 40:1769–1779

Carson D, Botha F (2002) Genes expressed in sugarcane maturing intermodal tissue. Plant Cell Rep 20:1075–1081

Casu RE, Grof CPL, Rae AL, McIntyre CL, Dimmock CM, Manners JM (2003) Identification of a novel sugar transporter homologue strongly expressed in maturing stem vascular tissues of sugarcane by expressed sequence tag and microarray analysis. Plant Mol Biol 52:371–386

Casu RE, Dimmock CM, Chapman SC, Grof CPL, McIntyre CL, Bonnett GD, Manners JM (2004) Identification of differentially expressed transcripts from maturing stem of sugarcane by in silico analysis of stem expressed sequence tags and gene expression profiling. Plant Mol Biol 54:503–517

Chepelev I, Wei G, Tang Q, Zhao K (2009) Detection of single nucleotide variations in expressed exons of the human genome using RNA-Seq. Nucleic Acids Res 37:e106

Cho RJ, Mindrinos M, Richards DR, Sapolsky RJ, Anderson M, Drenkard E, Dewdney J, Reuber TL, Stammers M, Federspiel N, Theologis A, Yang WH, Hubbell E, Au M, Chung EY, Lashkari D, Lemieux B, Dean C, Lipshutz RJ, Ausubel FM, Davis RW, Oefner PJ (1999) Genome-wide mapping with biallelic markers in *Arabidopsis thaliana*. Nat Genet 23(2):203–207

Cirulli ET, Singh A, Shianna KV, Ge D, Smith JP, Maia JM, Heinzen EL, Goedert JJ, Goldstein DB (2010) Screening the human exome: a comparison of whole genome and whole transcriptome sequencing. Genome Biol 11:R57

Cloonan N, Forrest A, Kolle G, Gardiner B, Faulkner G, Brown MK, Taylor DF, Steptoe AL, Wani S, Bethel G, Robertson AG, Perkins AC, Bruce SJ, Lee CC, Ranade SS, Peckham HE, Manning JM, McKernan KJ, Grimmond SM (2008) Stem cell transcriptome profiling via massive-scale mRNA sequencing. Nat Methods 5:613–619

D'Hont A, Lu YH, León DG, Grivet L, Feldmann P, Lanaud C, Glaszmann JC (1994) A molecular approach to unraveling the genetics of sugarcane, a complex polyploid of the Andropogoneae tribe. Genome 37(2):222–230

D'Hont A, Ison D, Alix K, Roux C, Glaszmann JC (1998) Determination of basic chromosome numbers in the genus *Saccharum* by physical mapping of ribosomal RNA genes. Genome 41:221–225

Da Silva JA, Bressiani JA (2005) Sucrose synthase molecular marker associated with sugar content in elite sugarcane progeny. Genet Mol Biol 28(2):294–298

Daugrois J, Grivet L, Roques D, Hoarau JY, Lombard H, Glaszmann JC, D'Hont A (1996) A putative major gene for rust resistance linked with a RFLP marker in sugarcane cultivar "R570". Theor Appl Genet 92(8):1059–1064

Elshire RJ, Glaubitz JC, Sun Q, Poland JA, Kawamoto K, Buckler ES, Mithcell SE (2011) A robust, simple genotyping-by-sequencing (GBS) approach for high diversity species. PLoS One 6(5):e19379

Garcia AAF, Kido EA, Meza AN, Souza HMB, Pinto LR, Pastina MM, Leite CS, Da Silva JAG, Ulian EC, Figueira AVO, Souza AP (2006) Development of an integrated genetic map of a sugarcane (*Saccharum spp.*) commercial cross based on a maximum-likelihood approach for estimation of linkage and linkage phases. Theor Appl Genet 112:298–314

Garcia AAF, Mollinari M, Marconi TG, Serang OR, Silva RR, Vieira MLC, Vicentini R, Costa EA, Mancini MC, MO G, Pastina MM, Gazaffi R, Martins ERF, Dahmer N, Sforça DA, Cardoso-Silva CB, Bundock P, Henry RJ, Souza GM, Van Sluys MA, Landell MGA, Carneiro MS, Vincentz MAG, Pinto LR, Vencovsky R, Souza AP (2013) SNP genotyping allows an in-depth characterisation of the genome of sugarcane and other complex autopolyploids. Sci Rep 3:3399

Gazaffi R, Margarido GRA, Pastina MM, Mollinari M, Garcia AAF (2014) A model for quantitative trait loci mapping, linkage phase, and segregation pattern estimation for a full sib progeny. Tree Genet Genomes 10:791–801

George AW and Aitken K (2010) A new approach for copy number estimation in polyploids. Jounal of Heredity 101:521–524

Grattapaglia D, Sederoff R (1994) Genetic linkage maps of eucalyptusgrandis and *Eucalyptus urophylla* using a pseudo-testcross: mapping strategy and RAPD markers. Genetics 137(4):1121–1137

Grivet L, Arruda P (2001) Sugarcane genomics: depicting the complex genome of an important tropical crop. Curr Opin Plant Biol 5:122–127

Grivet L, D'Hont A, Roques D, Feldmann P, Lanaud C, Glaszmann JC (1996) RFLP Mapping in Cultivated Sugarcane (*Saccharum* spp.): genome organization in a highly polyploid and aneuploid interspecific hybrid. Genetics 142:987–1000

Guimarães CT (1999) Mapeamento comparativo e detecção de QTLs em cana-de-açúcar utilizando marcadores moleculares. [s.l.] UFV, Viçosa, MG

Guimarães CT, Sills GR, Sobral BW (1997) Comparative mapping of Andropogoneae: *Saccharum* L. (sugarcane) and its relation to sorghum and maize. Proc Natl Acad Sci U S A 94(26):14261–14266

Gut IG (2004) DNA analysis by MALDI-TOF mass spectrometry. Hum Mutat 23(5):437–441

Hansey CN, Vaillancourt B, Sekhon RS, de Leon N, Kaeppler SM, Buell CR (2012) Maize (*Zea mays L.*) genome diversity as revealed by RNA-sequencing. PLoS One 7: e33071

Hoarau JY, Offmann B, D'Hont A, Risterucci AM, Roques D, Glaszmann GC, Grivet L (2001) Genetic dissection of a modern sugarcane cultivar (*Saccharum*spp.). I. Genome mapping with AFLP markers. Theor Appl Genet 103:84–97

Hoarau JY, Grivet L, Offmann B, Raboin LM, Diorflar JP, Payet J, Hellmann M, D'Hont A, Glaszmann JC (2002) Genetic dissection of a modern sugarcane cultivar (*Saccharum* spp.). II. Detection of QTLs for yield components. Theor Appl Genet 105:1027–1037

Huang Y-F, Poland JA, Wight CP, Jackson EW, Tinker NA (2014) Using genotyping-by-sequencing (GBS) for genomic discovery in cultivated oat. PLoS One 9(7):e102448

Irvine JE (1999) *Saccharum* species as horticultural classes. Theor Appl Genet 98:186–194

Jordan DR, Casu RE, Besse P, Carroll BC, Berding N, McIntyre CL (2004) Markers associated with stalk number and suckering in sugarcane colocate with tillering and rhizomatousness QTLs in sorghum. Genome 47:988–993

Kanazin V, Talbert H, See D, De Camp P, Nevo E, Blake T (2002) Discovery and assay of single-nucleotide polymorphisms in barley (*Hordeumvulgare*). Plant Mol Biol 48:529–537

Lai Z, Livingstone K, Zou Y, Church SA, Knapp SJ, Andrews J, Rieseberg LH (2005) Identification and mapping of SNPs from ESTs in sunflower. Theor Appl Genet 111:1532–1544

Lima MLA, Garcia AAF, Oliveira KM, Matsuoka S, Arizono H, Souza Junior CL, AP S (2002) Analysis of genetic similarity detected by AFLP and coefficient of parentage among genotypes of sugar cane (Saccharum spp.). Theor Appl Genet 104:30–38

Lister R, O'Malley RC, Tonti-Filippini J, Gregory BD, Berry CC, Millar AH, Ecker JR (2008) Highly integrated single-base resolution maps of the epigenome in Arabidopsis. Cell 133:523–536

Lu T, Lu G, Fan D, Zhu C, Li W, Zhao Q, Feng Q, Zhao Y, Guo Y, Li W, Huang X, Han B (2010) Function annotation of the rice transcriptome at single-nucleotide resolution by RNA-seq. Genome Res 20:1238–1249

Ma H-M, Schulze S, Lee S, Yang M, Mirkov E, Irvine J, Moore P, Paterson A (2004) An EST survey of the sugarcane transcriptome. Theor Appl Genet 108:851–863

Mantello CC, Cardoso-Silva CB, da Silva CC, de Souza LM, Scaloppi Junior EJ, de Souza GP, Vicentini R, Souza AP (2014) De novo assembly and transcriptome analysis of the rubber tree (*Hevea brasiliensis*) and SNP markers development for rubber biosynthesis pathways. PLoS One 9(7):e102665. doi:10.1371/journal.pone.0102665

Marconi TG, Costa EA, Miranda HRCAN, Mancini MC, Cardoso-Silva CB, Oliveira KM, Pinto LR, Molinari M, Garcia AAF, Souza AP (2011) Functional markers for gene mapping and genetic diversity studies in sugarcane. BMC Res Notes 4:264

Margarido GRA, Souza AP, Garcia AAF (2007) OneMap: software for genetic mapping in out-crossing species. Hereditas 144:78–79

Margarido GRA, Pastina MM, Souza AP, Garcia AAF (2015) Multi-trait multi-environment quantitative trait loci mapping for a sugarcane commercial cross provides insights on the inheritance of important traits. Mol Breed 35:175

McIntyre CL, Casu RE, Drenth J, Knight D, Whan VA, Croft BJ, Jordan DR, Manners JM (2005a) Resistance gene analogues in sugarcane and sorghum and their association with quantitative trait loci for rust resistance. Genome 48:391–400

McIntyre CL, Whan VA, Croft B, Magarey R, Smith GR (2005b) Identification and validation of molecular markers associated with Pachymetra root rot and brown rust resistance in sugarcane using map- and association-based approaches. Mol Breed 16:151–161

McIntyre CL, Jackson M, Cordeiro GM, Amouyal O, Hermann S, Aitken KS, Eliott F, Henry RJ, Casu RE, Bonnett GD (2006) The identification and characterisation of alleles of sucrose phosphate synthase gene family III in sugarcane. Mol Breed 18:39–50

Metzgar D, Jeffrey B, Wills C (2000) selection against frameshift mutations limits microsatellite expansion in coding DNA. Genome Res 10:72–80

Ming R, Liu SC, Lin YR, da Silva J, Wilson W, Braga D, van Deynze A, Wenslaff TF, Wu KK, Moore PH, Burnquist W, Sorrells ME, Irvine JE, Paterson AH (1998) Detailed alignment of

Saccharum and Sorghum chromosomes: comparative organization of closely related diploid and polyploid genomes. Genetics 150:1663–1682

Ming R, Liu SC, Moore PH, Irvine JE, Paterson AH (2001) QTL Analysis in a complex autopolyploid: genetic control of sugar content in sugarcane. Genome Res 11:2075–2084

Ming R, Liu SC, Bowers JE, Moore PH, Irvine JE, Paterson AH (2002a) Construction of a Saccharum consensus genetic map from two interspecific crosses. Crop Sci 42:570–583

Ming R, Wang W, Draye X, Moore H, Irvine E, Paterson H (2002b) Molecular dissection of complex traits in autopolyploids: mapping QTLs affecting sugar yield and related traits in sugarcane. Theor Appl Genet 105:332–345

Ming R, Del Monte TA, Hernandez E, Moore PH, Irvine JE, Paterson AH (2002c) Comparative analysis of QTLs affecting plant height and flowering among closely-related diploid and polyploid genomes. Genome 45:794–803

Mogg R, Batley J, Hanley S, Edwards D, O'Sullivan H, Edwards J (2002) Characterization of the flanking regions of *Zea mays* microsatellites reveals a large number of useful sequence polymorphisms. Theor Appl Genet 105:532–543

Morgante M, Hanafey M, Powell W (2002) Microsatellites are preferentially present with nonrepetitive DNA in plant genomes. Nat Genet 30:194–200

Morin R, O'Connor M, Griffith M, Kuchenbauer F, Delaney A et al (2008) Application of massively parallel sequencing to microRNA profiling and discovery in human embryonic stem cells. Genome Res 18:610–621

Morozova O, Marra M (2008) Applications of next-generation sequencing technologies in functional genomics. Genomics 92:255–264

Mudge J, Andersen WR, Kehrer RL, Fairbanks DJ (1996) A RAPD genetic map of *Saccharum officinarum*. Crop Sci 36:1362–1366

Nasu S, Suzuki J, Ohta R, Hasegawa K, Yui R, Kitazawa N, Monna L, Minobe Y (2002) Search for and analysis of single nucleotide polymorphisms (SNPS) in rice (*Oryza sativa*, Oryza rufipogon) and establishment of SNP markers. DNA Res 9:163–171

Oberacher H (2008) On the use of different mass spectrometric techniques for characterization of sequence variability in genomic DNA. Anal Bioanal Chem 391:135–149

Oliveira KM, Pinto LR, Marconi TG, Margarido GRA, Pastina MM, Teixeira LHM, Figueira AM, Ulian EC, Garcia AAF, Souza AP (2007) Functional genetic linkage map on EST-markers for a sugarcane (*Saccharum* spp.) commercial cross. Mol Breed 20:189–208

Oliveira KM, Pinto LR, Marconi TG, Mollinari M, Ulian EC, Chabregas SM, Falco MC, Burnquist AAF, Garcia AAF, Souza AP (2009) Characterization of new polymorphic functional markers for sugarcane. Genome 52:191–209

Palhares AC, Rodrigues-Morais TB, Van-Sluys MA, Domingues DS, Maccheroni W, Jordão H, Souza AP, Marconi TG, Mollinari M, Gazaffi R, Garcia AAF, Vieira MLV (2012) A novel linkage map of sugarcane with evidence for clustering of retrotransposon-based markers. BMC Genet 13:51

Pastina MM, Malosetti M, Gazaffi R, Mollinari M, Margarido GRA, Oliveira KM, Pinto LR, Souza AP, van Eeuwijk FA, Garcia AAF (2012) A mixed model QTL analysis for sugarcane multiple-harvest-location trial data. Theor Appl Genet 124:835–849

Pinto LR, Oliveira KM, Ulian EC, Garcia AAF, Souza AP (2004) Survey in the sugarcane expressed sequence tag database (SUCEST) for simple sequence repeats. Genome 47:795–804

Pinto LR, Oliveira KM, Marconi T, Garcia AAF, Ulian EC, Souza AP (2006) Characterization of novel sugarcane expressed sequence tag microsatellites and their comparison with genomics SSRs. Plant Breed 125:378–384

Pinto LR, Garcia AAF, Pastina MM, Teixeira LHM, Bressiani JA, Ulian EC, Bidoia MAP, Souza AP (2010) Analysis of genomic and functional RFLP derived markers associated with sucrose content, fiber and yield QTLs in a sugarcane (*Saccharum spp.*) commercial cross. Euphytica 172:313–327

Piperidis N, Jackson PA, D'Hont A, Besse P, Hoarau JY, Courtois B, Aitken KS, McIntyre CL (2008) Comparative genetics in sugarcane enables structured map enhancement and validation of marker-trait associations. Mol Breed 21:233–247

Poland J, Endelman J, Dawson J, Rutkoski J, Wu S, Manese Y, Dreisigackere S, Crossae J, Sánchez-Villedae H, Sorrellsd M, Jannink JL (2012) Genomic selection in wheat breeding using genotyping-by-sequencing. Plant Gen 5:103–113

Porceddu A, Albertini E, Barcaccia G, Falistocco E, Falcinelli M (2002) Linkage mapping in apomictic and sexual Kentucky bluegrass (*Poapratensis* L.) genotypes using a two way pseudo-testcross strategy based on AFLP and SAMPL markers. Theor Appl Genet 104:273–280

Raboin LM, Oliveira KM, Lecunff L, Telismart H, Roques D, Butterfield M, Hoarau JY, D'Hont A (2006) Genetic mapping in sugarcane, a high polyploid, using bi-parental progeny: identification of a gene controlling stalk colour and a new rust resistance gene. Theor Appl Genet 112:1382–1391

Rafalski A (2002) Applications of single nucleotide polymorphisms in crop genetics. Curr Opin Plant Biol 5:94–100

Rafalski A, Tingey S (2008) SNPs and their use in maize. In: Plant genotyping II: SNP [s.l.]. CAB International. pp 30–43

Reffay N, Jackson PA, Aitken KS, Hoarau JY, D'Hont A, Besse P, McIntyre CL (2005) Characterisation of genome regions incorporated from an important wild relative into Australian sugarcane. Mol Breed 15:367–381

Remington DL, Thornsberry JM, Matsuoka Y, Wilson LM, Whitt SR, Doebley J, Kresovich S, Goodman MM, Buckler ES (2001) Structure of linkage disequilibrium and phenotypic associations in the maize genome. Proc Natl Acad Sci U S A 98:11479–11484

Rostoks N, Mudie S, Cardle L, Russell J, Ramsay L, Booth A, Svensson JT, Wanamaker SI, Walia H, Rodriguez EM, Hedley PE, Liu H, Morris J, Close TJ, Marshall DF, Waugh R (2005) Genome-wide SNP discovery and linkage analysis in barley based on genes responsive to abiotic stress. Mol Genet Genomics 274:515–527

Salvi S, Sponza G, Morgante M, Tomes D, Niu X, Fengler KA, Meeley R, Ananiev EV, Svitashev S, Bruggemann E, Li B, Hainey CF, Radovic S, Zaina G, Rafalski JA, Tingey SV, Miao GH, Phillips RL, Tuberosa R (2007) Conserved noncoding genomic sequences associated with a flowering-time quantitative trait locus in maize. Proc Natl Acad Sci U S A 104:11376–11381

Santos FRC, Pinto LR, Carlini-Garcia LA, Gazaffi R, Mancini MC, Gonçalves BS, Medeiros CNF, Perecin D, Garcia AAF, Souza AP, Zucchi MI (2014) Marker-trait association and epistasis for Brown rust resistance in sugarcane. Euphytica 203:533–547

Serang O, Mollinari M, Garcia AAF (2012) Efficient exact maximum a posteriori computation for Bayesian SNP genotyping in polyploids. PLoS One 7(2):e30906. doi:10.1371/journal.pone.0030906

Setta N, Monteiro-Vitorello CB, Metcalfe CJ, Cruz GMQ, Del-Bem LE, Vicentini R, Nogueira FTS, Campos RA, Nunes SL, Turrini PCG, Vieira AP, Cruz EAO, Correa TCS, Hotta CT, Varani AAM, Vautrin S, Trindade AS, Vilela MM, Lembke CG, Sato PM, Andrade RF, Nishiyama-Jr MY, Cardoso-Silva CB, Scortecci KC, Garcia AAF, Carneiro MS, Kim C, Paterson AH, Bergès H, D'Hont A, Souza AP, Souza GM, Vincentz M, Kitajima JP, Van-Sluys M (2014) Building the sugarcane genome for biotechnology and identifying evolutionary trends. BMC Genomics 15(540):1–17

Sequenom (2007) Typer 4.0 Manual. Sequenom, San Diego, p. 179

Shepherd M, Cross M,·Dieters MJ, Henry R (2003) Genetic maps for Pinuselliottii var. elliottii and P. caribaea var. hondurensis using AFLP and microsatellite markers. Theor Appl Genet 106:1409–1419

Sills GR, Bridges W, Al-Janabi SM, Sobral BWS (1995) Genetic analysis of agronomic traits in a cross between sugarcane (*Saccharum officinarum* L.) and its presumed progenitor (*S. robustum* Brandes & Jesw. ex Grassl). Mol Breed 1:355–363

Silva JA, Sorrells ME, Burnquist WL, Tanksley SD (1993) RFLP linkage map and genome analysis of *Saccharum spontaneum*. Genome 36:782–791

Silva J, Honeycutt RJ, Burnquist W, Al-Janabi SM, Sorrells ME, Tanksley SD, Sobral BWS (1995) *Saccharum spontaneum* L. "SES 208" genetic linkage map combining RFLP and PCR-based markers. Mol Breed 1:165–179

Sonah H, Deshmukh RK, Sharma A, Singh VP, Gupta DK, Gacche RN, Rana JC, Singh NK, Sharma TR (2011) Genome-wide distribution and organization of microsatellites in plants: An insight into marker development in brachypodium. PLoS One 6(6):e21298

Tautz D (1989) Hypervariability of simple sequences as a general source of polymorphic DNA markers. Nucleic Acids Res 17:6463–6471

Tenaillon MI, Sawkins MC, Long AD, Gaut RL, Doebley JF, Gaut BS (2001) Patterns of DNA sequence polymorphism along chromosome 1 of maize (*Zea mays* ssp. mays L.). Proc Natl Acad Sci U S A 98:9161–9166

Toledo-Silva G, Cardoso-Silva CB, Jank L, Souza AP (2013) De novo transcriptome assembly for the tropical grass *Panicum maximum* Jacq. PLoS One 8(7):e70781. doi:10.1371/journal.pone.0070781

Toth G, Gaspari Z, Jurka J (2000) Microsatellites in different eukaryotic genomes: survey and analysis. Genome Res 10:967–981

Trick M, Long Y, Meng J, Bancroft I (2009) Single nucleotide polymorphism (SNP) discovery in the polyploid Brassica napus using Solexa transcriptome sequencing. Plant Biotechnol J 7:334–346

Van Belleghem SM, Roelofs D, Van Houdt J, Hendrickx F (2012) De novo TranscriptomeAssemb and SNP discovery in the wing polymorphic salt marsh beetle *Pogonuschalceus* (Coleoptera, Carabidae). PLoS One 7(8):e42605. doi:10.1371/journal.pone.0042605

Vettore AL, Silva FR, Kemper EL, Arruda P (2001) The libraries that made SUCEST. Genet Mol Biol 24:1–7

Vettore AL, da Silva FR, Kemper EL, Souza GM, da Silva AM, Ferro MI, Henrique-Silva F, Giglioti EA, MV L, LL C, Nobrega MP, Carrer H, França SC, BacciJúnior M, Goldman MH, Gomes SL, Nunes LR, Camargo LE, Siqueira WJ, Van Sluys MA, Thiemann OH, Kuramae EE, Santelli RV, Marino CL, Targon ML, Ferro JA, Silveira HC, Marini DC, Lemos EG, Monteiro-Vitorello CB, Tambor JH, Carraro DM, Roberto PG, Martins VG, Goldman GH, de Oliveira RC, Truffi D, Colombo CA, Rossi M, de Araujo PG, Sculaccio SA, Angella A, Lima MM, de Rosa Júnior VE, Siviero F, Coscrato VE, Machado MA, Grivet L, Di Mauro SM, Nobrega FG, Menck CF, Braga MD, Telles GP, Cara FA, Pedrosa G, Meidanis J, Arruda P (2003) Analysis and functional annotation of an expressed sequence tag collection for tropical crop sugarcane. Genome Res 13:2725–2735

Vos P, Hogers R, Bleeker M, Reijans M, Lee TV, Hornes M, Frijters A, Pot J, Peleman J, Kulper M, Zabeau M (1995) AFLP: a new technique for DNA fingerprinting. Nucleic Acids Res 23:4407–4414

Wei X, Jackson PA, McIntyre CL, Aitken KS, Croft B (2006) Associations between DNA markers and resistance to diseases in sugarcane and effects of population substructure. Theor Appl Genet 114:155–164

Williams JGK, Kubelik AR, Livak KJ, Rafalski JA, Tingey SV (1990) DNA polymorphisms amplified by arbitrary primers are useful as genetic markers. Nucleic Acids Res 18(22):6531–6535

Wilson LM, Whitt SR, Ibáñez AM, Rocheford TR, Goodman MM, Buckler ES (2004) Dissection of maize kernel composition and starch production by candidate gene association. Plant Cell 16:2719–2733

Wu RL, Han YF, Hu JJ, Fang JJ, Li L, Li ML, Zeng ZB (2000) An integrated genetic map of *Populusdeltoides* based on amplified fragment length polymorphisms. Theor Appl Genet 100:1249–1256

Wu R, Ma CX, Painter I, Zeng ZB (2002) Simultaneous maximum likelihood estimation of linkage and linkage phases in outcrossing species. Theor Popul Biol 61:349–363

Yu J, Hu S, Wang J, Wong GK, Li S, Liu B, Deng Y, Dai L, Zhou Y, Zhang X, Cao M, Liu J, Sun J, Tang J, Chen Y, Huang X, Lin W, Ye C, Tong W, Cong L, Geng J, Han Y, Li L, Li W, Hu G, Huang X, Li W, Li J, Liu Z, Li L, Liu J, Qi Q, Liu J, Li L, Li T, Wang X, Lu H, Wu T, Zhu M, Ni P, Han H, Dong W, Ren X, Feng X, Cui P, Li X, Wang H, Xu X, Zhai W, Xu Z, Zhang J, He S,

Zhang J, Xu J, Zhang K, Zheng X, Dong J, Zeng W, Tao L, Ye J, Tan J, Ren X, Chen X, He J, Liu D, Tian W, Tian C, Xia H, Bao Q, Li G, Gao H, Cao T, Wang J, Zhao W, Li P, Chen W, Wang X, Zhang Y, Hu J, Wang J, Liu S, Yang J, Zhang G, Xiong Y, Li Z, Mao L, Zhou C, Zhu Z, Chen R, Hao B, Zheng W, Chen S, Guo W, Li G, Liu S, Tao M, Wang J, Zhu L, Yuan L, Yang H (2002) A draft sequence of the rice genome (*Oryza sativa* L. ssp. *indica*). Science 296:79–92

Yu J, Buckler ES (2006) Genetic association mapping and genome organization of maize. Curr Opin Biotechnol 17:155–160

Part IV
Environmental and Policies Issues

Chapter 10
Environmental Sustainability Aspects of Second Generation Ethanol Production from Sugarcane

M.V. Galdos, H. Cantarella, A. Hastings, J. Hillier, and P. Smith

Abstract Sugarcane-derived ethanol from Brazil has a high output to input energy ratio and high greenhouse gas savings compared to fossil fuels. Current ethanol production is based on first generation (1G) technology, which ferments the sugars extracted from sugarcane stalks. Cellulosic ethanol (2G) can be produced from what is currently considered agricultural and agro-industrial residues (straw and bagasse), and also from dedicated high biomass-producing crops. 2G ethanol provides an opportunity to intensify production, obtaining more energy per unit of area cropped and potentially reducing the environmental footprint. Straw removal from the field has more negative than positive consequences for the environment and the production system, but the effects need to be evaluated at the site and regional levels. The use of bagasse as feedstock should be evaluated using life cycle assessment methods to account for its alternative uses as raw material and in cogeneration. Energy cane, a vigorous and rustic crop selected for total biomass production rather than for sucrose, is a promising feedstock for 2G ethanol. The presence of rhizomes, a deep root system, and intensive tillering contribute to erosion control, crop longevity, and soil carbon sequestration. There is a need for more long-term experiments focusing on soil quality, nutrient cycling, greenhouse gas emissions, and crop production using innovative techniques such as stable isotope labeling and

M.V. Galdos (✉)
School of Biosciences, University of Nottingham, Sutton Bonington LE12 5RA, UK

Interdisciplinary Center of Energy Planning (NIPE),
Unicamp, R. Cora Coralina 330, Campinas, SP 13083-896, Brazil
e-mail: marcelo.galdos@nottingham.ac.uk

H. Cantarella
Soil and Environmental Resources Center, Agronomic Institute (IAC),
Av. Barão de Itapura 1481, Campinas, SP 13013-970, Brazil

A. Hastings • J. Hillier • P. Smith
Institute of Biological and Environmental Sciences, University of Aberdeen,
23 St Machar Drive, Aberdeen AB24 3UU, UK

M.S. Buckeridge, A.P. De Souza (eds.), *Advances of Basic Science for Second Generation Bioethanol from Sugarcane*, DOI 10.1007/978-3-319-49826-3_10

intensive soil flux measurements with automatic chambers to understand the impact of removing crop residues for bioenergy production and of using high-biomass dedicated crops. Process-based models are useful in sustainability assessments of 2G sugarcane ethanol production since they take into account site and regional variability in soil biogeochemistry, climate parameters, management practices, plant genetic traits, and the interactions of these factors. The integration of models and geographic information systems allows for regional assessments of the potential impacts of bioenergy production, contributing to the identification and promotion of sustainable pathways for cellulosic ethanol production.

Keywords Crop residues • Energy cane • Modeling • Life cycle assessment

10.1 Introduction

Sugarcane (*Saccharum* spp.) is a C_4 plant highly efficient in turning solar radiation into biomass. In Brazil, sugarcane is used to produce ethanol, sugar, and electricity from cogeneration. Sugarcane-derived ethanol from Brazil has been considered an advanced biofuel, a better energy ration, and lower life cycle greenhouse gas emissions when compared to fossil fuels. Current ethanol production is based on first generation (1G) technology, fermenting sugars extracted from sugarcane stalks. However, sugar only represents approximately one-third of the energy content of sugarcane. The other two-thirds are composed of straw that is either burned on the field or left as mulch, and bagasse, the fibrous material left from the juice extraction process, which is mostly used as fuel for process heat and electricity generation at the mill.

Cellulosic ethanol, also known as second generation (2G) ethanol, can be produced from what is currently considered agricultural and industrial residues (straw and bagasse). With the development of commercially viable lignocellulosic ethanol technology and the possibility to use other parts of the plant as feedstock, new scientific challenges will emerge in assessing the sustainability of production systems. Besides using crop residues for energy production, new high yield crops are being evaluated for 2G ethanol production, which will also generate the need for new sustainability assessment methods and tools. Biomass production has environmental impacts on soil quality, greenhouse gas emissions, biodiversity, and water resources. One of the main challenges in sustainability science is to develop and apply methodologies to accurately estimate those impacts considering regional differences, to ultimately identify practices to mitigate this impact. In order to achieve this goal, it is necessary to combine innovative measurement techniques, process-based models, and spatially explicit assessments, as discussed in this chapter.

10.2 Crop Residues as Feedstock for 2G Ethanol

In the sugarcane production system, burning the residues has been a common practice to facilitate harvest and transport operations. Burning crop residues results in emissions of gases such as CO_2, CH_4, and N_2O, which contribute to the greenhouse effect. Burning also emits soot, which causes nuisance as well as possible health damage to the people living in the surroundings of the crops (Cançado et al. 2006). Black carbon is a major component of soot, resulting from the incomplete combustion of fossil fuels, biofuels, and biomass. Black carbon scatters and absorbs portions of incoming solar rays, but also absorbs radiation from the diffuse upward rays of scattered sunlight, with a net effect of warming of the atmosphere (Galdos et al. 2013).

Due to environmental and economic reasons, there is ongoing burning phase out programs in the main sugarcane-growing regions in Brazil, with the gradual replacement of manual harvest with burning by mechanized harvest without burning. The slope of terrain is one of the limiting factors for harvesting sugarcane mechanically. Currently, areas with a slope of 12% or less are considered mechanizable. In São Paulo state, the largest sugarcane producer in Brazil, preharvest burning is expected to cease in all areas suitable for mechanical harvest by 2021 by state law, but a voluntary sugarcane industry program (Environmental Protocol) has set even more aggressive targets for phasing out preharvest burning (Fig. 10.1). Most of the sugarcane area in the state is on gentle slopes, and thus suitable for mechanization. In the 2013/2014 crop season, 83% of the sugarcane area was harvested without burning in São Paulo state (UNICA 2015). The progressive prohibition of burning and the development of mechanical harvesters have led to an increase in green cane harvesting (harvesting without burning) leaving 8–20 tons (dry matter) of straw per hectare on the field, mainly comprised of dry leaves and tops (Franco et al. 2013;

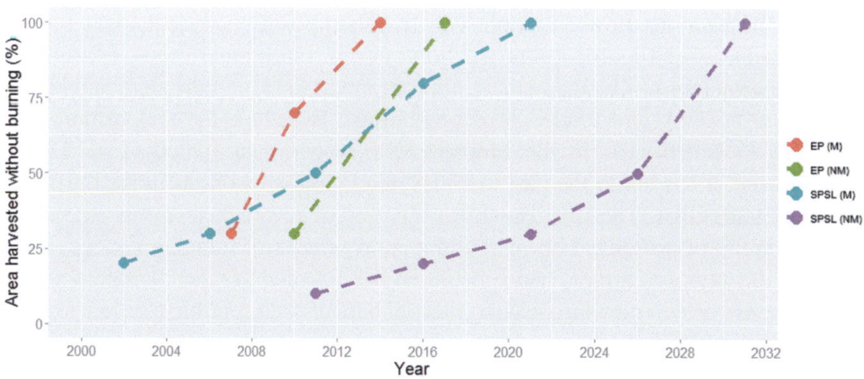

Fig. 10.1 Schedule for phasing out preharvest sugarcane burning in São Paulo state. SPSL (São Paulo State Law); EP (Environmental Protocol) in Mechanizable areas (M, slope ≤12%) and Non-mechanizable areas (NM, slope >12%)

Table 10.1 Consequences of straw removal for 2G on the agroecosystem

Effect of straw collection on	Type of impact
Soil organic C and C sequestration	Negative
Nutrient cycling	Negative; sometimes only long term effect
Sugarcane yield	Usually negative
Soil compaction and other physical properties	Usually negative
Sugarcane sprouting or ratoon regrowth	Usually negative
Soil erosion and nutrient losses	Highly negative
Weed control	Negative, but type of weed may change
Soil moisture	Negative
Greenhouse gases emission	Usually increases
Pests and diseases	Usually positive
Risk of fire	Positive
Ease of field operations	Positive

Horta Nogueira and Leal 2012; Leal et al. 2013; Trivelin et al. 2013). Although 2G bioethanol can be produced using only the bagasse that is transported to the mill, the collection of harvest residues will always be an issue because of the large quantity of energy it potentially contains—about 1/3 of the energy of the whole sugarcane plant—and of the economy of the whole process (Horta Nogueira and Leal 2012).

The impact of sugarcane straw removal on soil quality and, consequently, on the long-term sustainable plant production as well as the question of how much plant material can be collected with little or no impact was the object of a comprehensive study by Hassuani et al. (2005) and later on by several other authors (Cantarella et al. 2013; Leal et al. 2013; Sordi and Manechini 2013). The main results from these studies are summarized in Table 10.1. It is well established that the effect of straw preservation or removal is site specific (Marin et al. 2014; Thorburn et al. 2012) and, in many cases, the net effect cannot be clearly defined. However, in general, the straw removal has more negative than positive consequences for the environment as well as for the production system. For annual crops, continuous complete crop removal by using the grain for human food, and the stem and leaves for animal bedding and fodder have been shown to reduce the soil carbon, and therefore its fertility, as in the Morrow Plots experiment (Odell et al. 1982). Furthermore, it has led in the past to widespread wind erosion, as in the dustbowl event in 1930s USA. The Dust Bowl was a period of severe dust storms caused by a combination of prolonged drought and inappropriate soil tillage practices in the Great Plains region. Farmers had substituted deep-rooted grasses by annual crops using extensive deep plowing, leaving the soil bare and susceptible to wind erosion (Lee and Gill 2015).

Nutrient cycling in sugarcane is clearly improved with straw preservation, which in turn, reduces the fertilization needs. Rossetto et al. (2010), for example, showed a small response to potassium (K) in 15 fields where the straw was preserved. On the other hand, nitrogen (N) rates are likely to be increased because of the high C:N ratio and slow decomposition of the residue, which hardly contribute to sugarcane nutrition in the short run (Vitti et al. 2011). However, the continuous addition of straw will cause an increase in organic N pool. A new equilibrium is estimated to take 20–30 years (Robertson and Thorburn 2007) or even 40 years (Trivelin et al. 2013) and, in the long run, N fertilization may also be reduced. Franco et al. (2013) and Trivelin et al. (2013) showed that approximately 80% of the nutrients present in the straw are in the tops—the green part of the harvest residues—and suggested that only the dry leaves should be collected for bioenergy. In addition, the dry leaves presented higher yields in the pretreatment step for 2G production (Franco et al. 2013). Further, the separation of the tops with mechanical harvest is a challenge not only because presently the machines are not adapted to that, but also because plants in the field do not have a uniform height and high yielding sugarcanes tend to bend.

A thick mulch of straw may delay the sprouting of the ratoons and reduce plant stand of some sugarcane varieties. However, new sugarcane varieties are being tested under green cane, and it is likely that in the near future the plant material available will be adapted. Viator et al. (2006) observed that extracts of sugarcane straw presented autotoxic effect delaying early leaf development on ratoons, but this only happened under high extract concentrations; at low concentration the extracts stimulated sugarcane growth. In the unburned cane harvest system, a significant amount of straw on the ground can create an ideal microclimate (mainly temperature and humidity) for the development of weeds, pests, and diseases infestations (Wisniewski and Holtz 1997). Weeds, pests, and diseases present in sugarcane fields are important due to damage caused to stalks, tillers, leaves, root system, and stalk base, with larger infestations occurring, in general, in older cane (Hassuani et al. 2005). The incidence of several sugarcane pests is likely to increase under straw blanketing. However, the reference system is usually fields where sugarcane was burned. Thus, it is still unclear whether the straw collection will substantially decrease the pest population because fire is not present (Dinardo-Miranda and Fracasso 2013).

Many studies have shown that straw preservation significantly contributes to increasing soil carbon (C), and consequently soil organic matter which is highly correlated with soil quality (Cerri et al. 2013; De Luca et al. 2008; De Resende et al. 2006; Galdos et al. 2009a, b, 2010; Pinheiro et al. 2010). Several studies have shown that increased soil C sequestration under the green cane management system is affected by climate conditions (Robertson 2003), soil texture (Silver et al. 2000), time since green cane management implementation, and N fertilizer management (Graham et al. 2002a, b). Considering the large quantities of residue generated and their high carbon-to-nitrogen ratio and fiber content, it is likely that removing part of the straw will still secure most of the soil carbon sequestration potential. Nevertheless, the amount sustainably removed would be site specific, and should be calculated considering climate, topography, soil, and crop variables.

Straw removal may indirectly cause yield reduction (De Resende et al. 2006; Marin et al. 2014). In addition, increasing C sequestration in soils may be an important way of mitigating the climate warming effects of an excess of CO_2 concentration in the atmosphere (Lal 2013). However, there is a limit to which soils can store organic C (Galdos et al. 2009b; Lal 2008) so that after some time, the addition of organic material will have little effect on soil C accumulation although equilibrium is reached only in the long term. In fact, in some soils, studies spanning more than five decades have shown only small increases in the soil organic C stock from straw preservation, although other soil properties were positively affected (Graham et al. 2002a, b).

The contribution of roots to soil C is usually overlooked. Otto et al. (2009) found that the root system could reach 5–7 t ha^{-1} (dry matter) at the peak of the vegetative stage of a high yielding sugarcane; root dry matter was reduced to about 2.5 t ha^{-1} at the time of plant maturity for harvest. About 20–35% of the sugarcane plant dry matter may be allocated to the root system (Carvalho et al. 2013). Therefore, studies are needed to determine whether and in what situations the root system of sugarcane can supply C necessary for preserving or increasing soil organic matter content.

Some important roles of straw cover on the soil, such soil moisture preservation (Figueiredo and La Scala Jr. 2011), erosion control (Silva et al. 2012), and weed control (Hassuani et al. 2005), cannot be replaced by C input by roots. However, such beneficial effects can be achieved even when only part of the harvest residues are preserved, indicating that in many situations the collection of part of the straw for 2G ethanol production will not jeopardize soil quality and long-term yield potential. However, more research and modeling efforts are needed to help to define the adequate trade-off between using harvest residues for bioenergy and the preservation of their benefits if left on the field.

Long-term experiments are required to determine the impact on soil carbon and soil fertility of removing crop residues from the field. Such experiments will allow calculations of the optimum level of straw removal that will maximize the ethanol production while preserving soil carbon and fertility. As variation in input will be provided by C_4 sugarcane, experiments to determine the fate of soil carbon can be made by ^{14}C pulse labeling and measuring the partitioning of the soil organic carbon into light and heavy size fractions (Zimmerman et al. 2011). These types of experiments will also enable the parameterization of soil carbon turnover models for this crop management.

10.3 Energy Cane, Sweet Sorghum, and Miscanthus

Modern sugarcane cultivars are complex poly alloaneuploids comprising the genome of *Saccharum officinarum*, varying proportions of *Saccharum spontaneum* and recombinant or translocated chromosomes (Matsuoka et al. 2014—see Chap. 9). *S. officinarum* is a juicy sugary form, and *S. spontaneum* is a grassy wild species. Crop breeding for sugarcane has been directed at increasing sugar productivity, in detriment of fiber. Energy cane tough is a distinct form of the plant that was selected for total biomass production rather than for sucrose. Its development started in Louisiana and Puerto Rico in the USA in the late 1970s aiming at a more vigorous and rustic crop with a high yield of fiber (Alexander 1985).

Compared to sugarcane, energy cane cultivars are less demanding in nutrients, more resistant to diseases, and with higher competitive ability against weeds. Field studies in tropical and subtropical regions have shown high biomass productivity, with more than 250 t ha^{-1} $year^{-1}$ in rainfed conditions (close to 60 t ha^{-1} $year^{-1}$ dry matter) (Matsuoka et al. 2014). The production of a high number of tillers per stool, both in the plant and ratoon phases, is a characteristic of energy cane. For 2G ethanol production, the high total biomass yield implies a more efficient use of land, producing more energy per hectare, with clear positive implications to economic and environmental sustainability. Since energy cane is less demanding in water and nutrients, there is also high potential to use degraded or less fertile areas, thus avoiding direct competition with food (Carvalho-Neto et al. 2014).

Another relevant aspect of the energy cane is the more vigorous and deeper root system, as compared with sugarcane varieties (Fig. 10.2). Root systems are directly related to soil structure and water infiltration, thus helping to prevent erosion by runoff. Carbon input from root turnover is one of the main factors affecting soil carbon sequestration, so it is expected that a more vigorous root system will contribute to increasing carbon fixation in the soil in the long term. There is also evidence of a rhizomatous habit in energy cane (Fig. 10.3), not present in sugarcane. Rhizomes provide the ability to overcome abiotic stress conditions, such as water deficit, freezing temperatures, and soil compaction from traffic (Matsuoka et al. 2014). However, increased CO_2 concentrations in the future might negatively affect rhizome carbon accumulation. De Souza et al. (2013), in an experiment with miscanthus, described a decrease in starch content in roots and rhizomes with elevated CO_2 concentration under open-air conditions.

Fig. 10.2 Comparison between root systems of energy cane (**a**) and sugarcane (**b**) in 10-month-old plant crops grown in adjacent fields. Photo: Sizuo Matsuoka, Vignis S.A.

Fig. 10.3 Rhizomes in energy cane. Photo: Sizuo Matsuoka, Vignis S.A.

The intensive tillering, vigorous root systems, and the presence of rhizomes also contribute to increasing crop longevity. Sugarcane is a perennial grass that needs to be replanted every 6 years on average, due to yield decline. In preparation for the crop renewal, the stool of the last harvested crop is killed and the soil is prepared for the new crop. Under the widespread conventional cultivation system, the whole replanting area is disturbed using tillage, disking, and, sometimes, subsoiling. Part of the soil carbon stored under the green cane management can be lost during the replanting period (Resende et al. 2006). The positive correlation between soil tillage and increased soil C mineralization has been observed in several studies, being attributed mainly to the destruction of soil aggregates and the subsequent exposure of the organic material to decomposing biota (Paustian et al. 1997). The longevity of energy cane is higher than sugarcane, reaching 10 or more years in a cycle before having to be replanted. There are in the long run, therefore, fewer soil disturbance events, with implications on both the energy use in the agricultural phase of ethanol production—lower diesel use from cultivation events—and in the potential for soil carbon sequestration. Since the replanting period leaves the soil bare until a new crop is planted, fewer cultivation events in the long term also mean lower exposure to water and wind erosion.

Sweet sorghum (*Sorghum bicolor* L. Moench) has been evaluated as a feedstock for both 1G and 2G ethanol production (Ratnavathi et al. 2011). Currently, there is preliminary research in Brazil for cropping sweet sorghum in the short period between sugarcane cycles, providing supplementary feedstock for ethanol mills in a period of low sugarcane availability. Experiments with another 2G bioenergy feedstock in Europe, C_4 *Miscanthus x giganteus*, has enabled the quantify and parameterize soil/crop models related to the impact of land use change from grassland and arable land to perennial miscanthus grasses. This has been achieved by investigating the fate of miscanthus organic material input and the soil carbon stored from the previous land use. This has been relatively easy to achieve by measuring the changes in carbon isotope ratios as the miscanthus carbon from C_4 photosynthesis is distinct from the C_3 plants, which include the native European grasses and other crops. These experiments have shown that arable land conversion results in a soil carbon increase, and changes from grasses to miscanthus plantations result in no overall change (Dondini et al. 2009; Zatta et al. 2013). In the USA, experiments and Life Cycle Assessments focusing on miscanthus have identified high potential for carbon sequestration in roots and the soil, and reduced the need for nutrient additions (Zeri et al. 2011; Masters et al. 2016). However, the agronomy of miscanthus is different from sugarcane and energy cane. Miscanthus is allowed to senesce in the field, and the nutrients are repartitioned to the rhizomes before harvest, while sugarcane and energy cane are cut green. This means that both root turnover and leaf fall are part of the miscanthus soil input and can be up to one-third of the biomass harvested.

10.4 Greenhouse Gas Emissions

Decreasing or maintaining low greenhouse gases (GHGs) emissions in all phases of the production cycle is a crucial condition for sustainable bioenergy. The effect of straw on GHGs emission from the soil system is complex. Signor et al. (2014) observed that removing straw from the field did not change GHG emissions as compared to preserving the plant blanketing. However, Figueiredo et al. (2014) showed that when straw was removed there was an increase in CO_2 emission from soil organic C. In another study, Carmo et al. (2013) tested the effect of varying amounts of sugarcane straw on N_2O emission from N fertilizer and observed that emissions were weakly affected by plant material when the straw rate was up to 15 t ha^{-1} dry matter, but increased substantially at 20 t ha^{-1} (Fig. 10.4), an amount of straw that is not usual under the present yield levels. Soil moisture seems to interfere with the interaction of straw and N fertilizer that controls N_2O emission. Vargas et al. (2014) found that under low moisture the amount of straw did not affect N_2O emission, but under high moisture, the presence of straw increased two-fold the release of this GHG.

Vinasse, a by-product of ethanol production, is recycled in the sugarcane fields supplying nutrients and organic C. As vinasse contains N, its application to soil increases N_2O emissions, and these are slightly higher in soils without straw (Oliveira et al. 2013). However, this study showed that the emission factor (how much N is lost as N_2O compared to how much is applied) associated with the N contained in vinasse (0.44–0.68%) is lower than the default value of 1% used in the models by the International Panel of Climate Changes (IPCC). Similarly, Siqueira

Neto et al. (2015) calculated an emission factor of 0.65% for vinasse application, considering an application rate of 150 m^3 per hectare. However, Paredes et al. (2014) suggested that the average factor for vinasse in several Brazilian studies is 1.94%.

The application of vinasse and N fertilizer on a sugarcane field with varying amounts of straw resulted in high N_2O emissions (Carmo et al. 2013). When vinasse was added soon after N fertilization, the N-N_2O emission factor of the fertilizer increased by about 12% per ton of straw and reached an emission factor of 3.0% at the highest rate of straw, 20 t ha^{-1} (Fig. 10.4). Vinasse supplies readily available C in addition to increasing soil moisture, which explains the high N_2O emissions. As both vinasse and N fertilizer are common inputs in sugarcane fields, an alternative to reduce emissions is to apply them separately on different occasions and locations in the field (Carmo et al. 2013).

The vinasses used in the published studies are from sugarcane juices or molasses. There is little information about the composition of vinasse from lignocellulosic materials because 2G ethanol is not being produced in large scale so far. Its composition will likely vary with the feedstock. Vinasse from sugarcane bagasse has approximately the same content of organic C as juice vinasse, but the nutrient content, especially of K, is much lower (Moraes et al. 2015). Although the experience so far with first generation vinasse will be useful, the management of 2G vinasse and the consequences of its use in the field remain to be studied.

10.5 Process-Based Modeling

Modeling can be an important tool in understanding carbon and nutrient dynamics in agricultural systems. By using models, it is possible to evaluate the long-term effects of different residue management practices considering site-specific conditions such as soil texture, soil water dynamics, and different agricultural input types

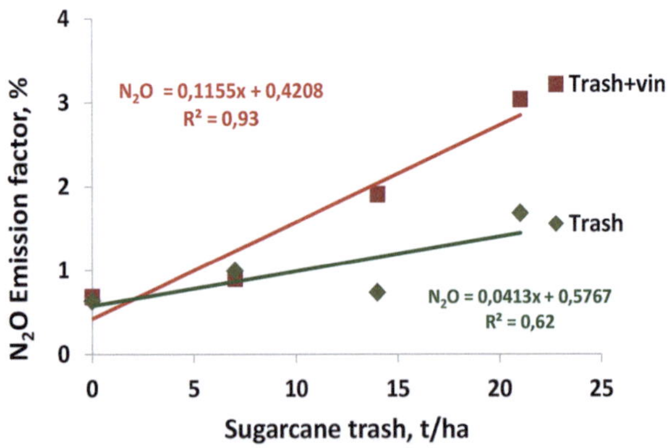

Fig. 10.4 Effect of trash and vinasse on N_2O emission factor of N fertilizer in a sugarcane field. From Carmo et al. (2013)

and rates. Models can be useful in evaluating sustainability aspects of both crop removal and the use of new feedstocks for 2G ethanol production. Biogeochemical models describe the main processes involving carbon, nutrients, water, and emissions in the soil/plant/atmosphere interface, as illustrated in Fig. 10.5 for the ECOSSE model.

Muth et al. (2012) described a methodology to integrate soil physical and chemical parameters, and yield data and slope. This methodology allows the evaluation in a spatially explicit manner how much crop residue can be sustainably removed considering factors such as soil carbon stocks, soil erosion, and crop productivity. Duval et al. (2013) used the DayCent biogeochemical model (Fig. 10.6) to simulate energy cane production in central Florida (USA) in two different soil types including emissions from the expansion into pastureland. The authors concluded that energy cane grown on nutrient poor soils has the potential to be a high-yielding 2G ethanol feedstock, and it acts as a GHG sink.

The issue of land use change has been at the forefront of the discussion on sustainability of bioenergy. The direct and indirect environmental impacts of land use change related to bioenergy potentially include deforestation, increased greenhouse gas emissions, soil and water degradation, biodiversity losses, and competition with food production. Land use change can produce significant GHG emissions from biomass burning and decomposition of soil organic matter, especially when vegetation with high carbon stocks, such as forests, are replaced by crops (Buckeridge et al. 2012). Deforestation caused directly or indirectly by bioenergy crop expansion can deny all or part of the GHG mitigation benefits of biofuels (Miyake et al. 2012).

The integration of process-based models and geographic information systems (GIS) allows for regional assessments of the potential impacts of bioenergy crop expansion on water resources, carbon dynamics, and biodiversity. In a study focusing on soybean and corn expansion in North Dakota using a land use change model and a recharge estimation model—including climate change scenarios—Li and Merchant (2013) identified areas with high vulnerability for groundwater contamination from nitrates. This would be caused primarily by the displacement of crops such as wheat and alfalfa, which have lower fertilizer application rates than corn. Although there is limited land available for bioenergy production in Europe, spatially explicit estimates of environmental impacts of the direct land use change occurring within the European Union states reveal that depending solely on 1G biofuels to fulfil bioenergy targets would lead to unintended negative consequences (Humpenöder et al. 2013). However, Hastings et al. (2009) showed a positive impact on soil carbon of growing miscanthus as a 2G crop on excess arable land in the EU, and Don et al. (2012) comparing 1G and 2G crops in Europe had the same conclusion.

Focusing on potential impacts of crop residue removal on soil carbon stock depletion, Lugato and Jones (2014) used an integration of the CENTURY model and GIS to evaluate the impact of low (30%) and high (90%) maize stover removal rates for 2G ethanol production at the European level. In the high removal scenario, mitigation practices such as using a ryegrass cover crop and returning biodigestates to the soil were considered in the modeling. The authors concluded that there would be soil carbon depletion of 39.7 Mt. CO_2 and 135.4 Mt. $CO_{2\,eq}$ by 2020 in the low

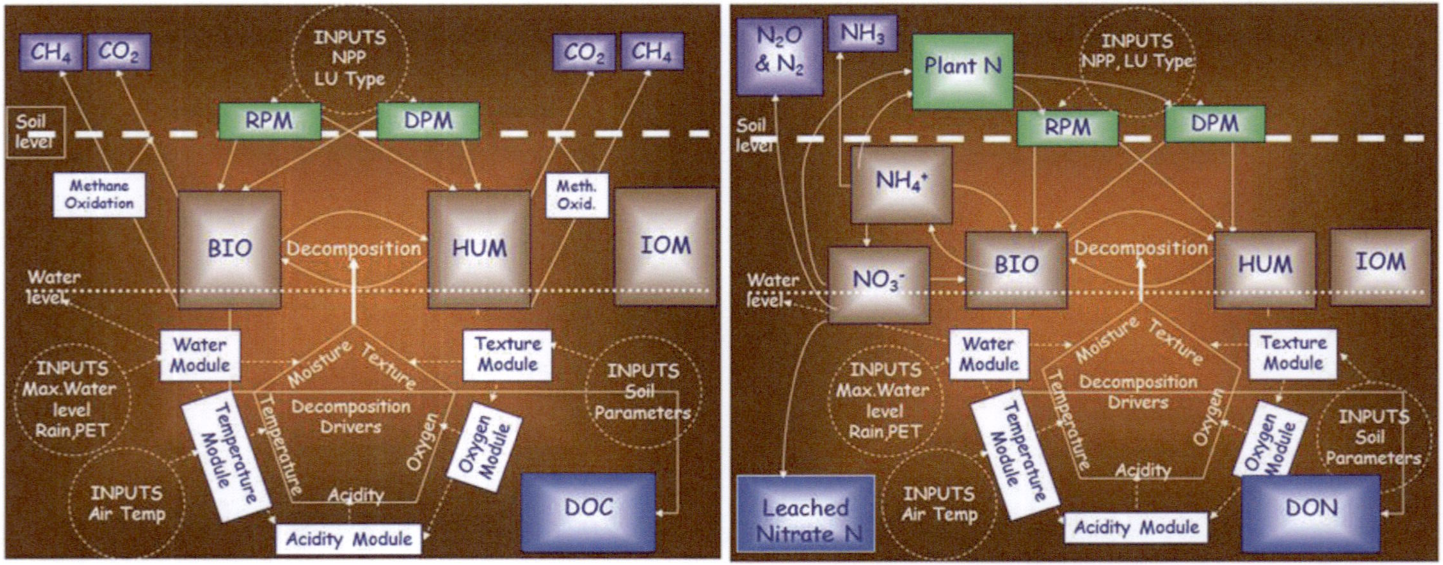

Fig. 10.5 Structure of the **(a)** carbon and **(b)** nitrogen components of ECOSSE as an example of a process-based model. Plant inputs enter the soil as RPM (resistant plant material) and DPM (decomposable plant material), and decompose into BIO ('biomass' or active organic matter) and HUM ('humus' or more slowly turning over soil organic matter). Organic matter that has become inert (IOM) is assumed to not contribute to the decomposition processes. Losses of C and N from the soil are gaseous (CH_4, CO_2, N_2O, N_2, and NH_3) and in solution (*DOC* dissolved organic C, *DON* dissolved organic N, and leached nitrate N). *Solid arrows* indicate flow of material. *Dashed arrows* indicate influence. After Smith et al. (2010)

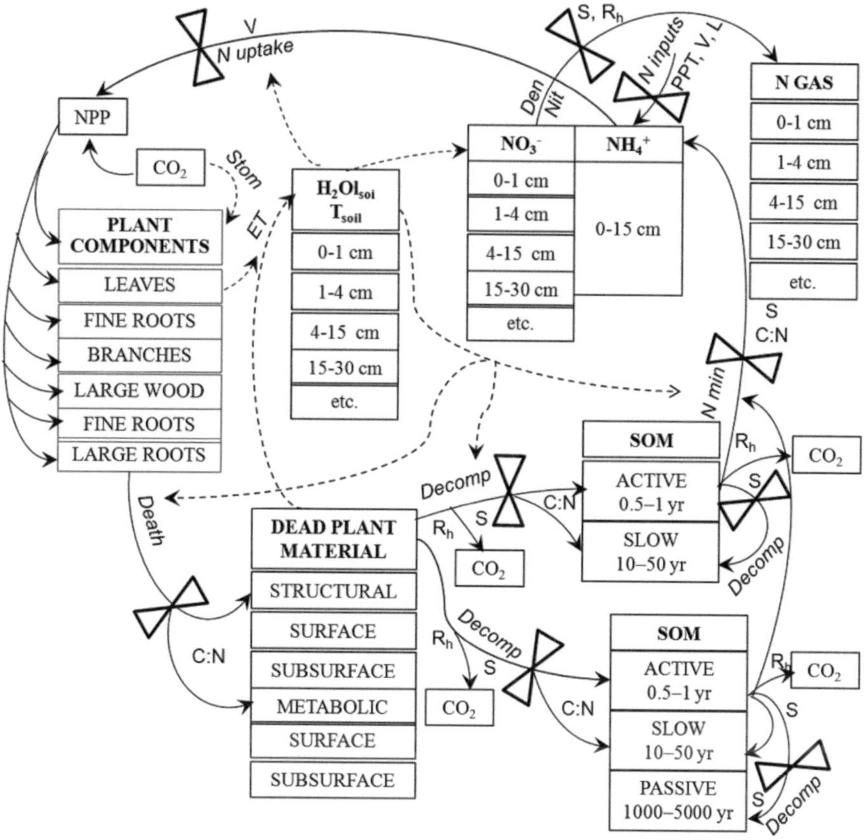

Fig. 10.6 Special features of the DayCent modeling package, including carbon and nitrogen stocks and fluxes and the main crop, soil and atmosphere pools (Del grosso et al. 2011)

and high stover removal rates, respectively. Mitigation practices would diminish soil carbon losses, but would not totally offset the negative soil C balance.

Models can also be used to evaluate scenarios for the implementation of alternative feedstocks for 2G ethanol production at the regional scale. Davis et al. (2012) used the DayCent model to estimate the effects on ecosystem services of replacing corn ethanol feedstocks with the perennial cellulosic feedstocks switchgrass (*Panicum virgatum* L.) and miscanthus (*Miscanthus x giganteus* Greef et Deuter) at a county-level resolution for the central US region. The modeling results of replacing corn by those cellulosic crops indicate a potential increase in ethanol and grain for food production and a reduction in nitrogen leaching and greenhouse gas emissions.

Life Cycle Assessments (LCAs) of current sugarcane ethanol production include the use of bagasse as a fuel and the export of electricity to reach the low GHG emissions level and high energy balance. Nevertheless, these numbers will change if the straw and the bagasse are used for the production of 2G ethanol so that a comparison of LCAs with and without the 2G ethanol production is required to ensure that this is not a retrograde step in reducing overall GHG emissions. Also, some bagasse is used for the production of paper, cardboard, and construction boards and to provide heat and power for other industries. The impact of their substitution by other materials and fuels will also need to be considered in the overall LCA and the emissions that will be generated by the use of alternative material and processes.

Dias et al. (2013) performed an LCA of sugarcane ethanol in Brazil considering three scenarios: (1) conventional first generation (1G), (2) stand-alone second generation (2G), and (3) an integration of first and second generation (1G + 2G). The authors concluded that integration of first and second generation ethanol production processes could be more economical, more efficient, and present lower environmental impacts than the first generation or the stand-alone second generation, mainly since the output of energy per unit of feedstock used would be maximized. Sheehan et al. (2003) built a life cycle model to simulate the collection of corn stover in the state of Iowa (USA) to produce ethanol, including impacts on soil carbon dynamics, soil erosion, and yield. For Iowa cornfield conditions, modeling results indicated that the minimum amount of residues that should be left on the field was, on average, 4.9 and 2.5 t ha^{-1} for conventional tillage and no-till, respectively. Life cycle assessments of 2G ethanol production are also very sensitive to the allocation methodologies applied (e.g., mass, energy, economic), which are related to how crop and agro-industrial residues such as straw and bagasse are accounted for.

10.6 Conclusion and Perspectives

Lignocellulosic ethanol production provides an opportunity to intensify production, obtaining more energy per unit of area cropped with sugarcane, by enabling the use of crop residues and highly productive fiber sources such as energy cane as feedstock. If agricultural inputs remain the same, the increase in energy output per unit of area or unit of feedstock will imply an increase in production efficiency and a smaller environmental footprint. Producing more energy per unit of land avoids land use change, which has direct and indirect effects on greenhouse gas emissions, soil and water quality, and biodiversity.

There are clear benefits of leaving crop residues on the field, such as protection against erosion, nutrient cycling, soil carbon sequestration, weed suppression, and soil moisture retention. On the other hand, considering the large quantities of residue generated and their high carbon-to-nitrogen ratio and fiber content, it is likely that removing part of the straw will still secure most environmental and agronomic benefits. Those benefits are site specific so that the amount of straw that can be removed sustainably should be calculated considering climate, topography, soil, and crop variables. Best management practices such as no till, unburned harvest, crop rotation with N fixing crops, extending the plant/ratoon cycle, and leaving at least part of the straw on the soil surface are important to secure the sustainability of second generation sugarcane ethanol.

More long-term experiments focusing on soil quality, nutrient cycling, GHG emissions, and crop production are necessary to understand the impact of removing crop residues for bioenergy production and of using high-biomass dedicated crops. Novel approaches and metrics should be used in assessing the relative impacts of agricultural management changes related to cellulosic ethanol production. For instance, stable isotopes can be used to assess the relative contribution of sugarcane surface litter and roots to soil carbon dynamics through $\partial^{13}C$ labeling, and nutrient cycling can be assessed using ^{15}N labeled fertilizer and straw. Regarding soil GHG emissions, due to the high spatial and temporal variability, intensive soil flux measurements in automated chambers should be established to understand the impact of crop residue removal and new feedstocks on soil carbon and nutrient stocks and fluxes.

Sustainability aspects of 2G sugarcane ethanol production need to take into account local and regional variability in soil biogeochemistry, climate parameters, management practices, plant genetic traits, and the interactions of these factors. Calibrated and validated models can be very useful for simulating processes in the soil/plant/atmosphere interface at the field scale. Furthermore, geographic information systems containing soil, climate, and land use databases at the regional scale can be integrated with process-based models to upscale simulations and test the regional impacts of different bioenergy production scenarios and to identify sustainable production pathways.

Acknowledgements The authors wish to acknowledge Dr. Sizuo Matsuoka and Dr. José Bressiani for the support with information on energy cane. This work was supported by grant 2012/06933-6 from the São Paulo Research Foundation (FAPESP) and the Brazil Partnering Award of the Biotechnology and Biological Science Research Council (BBSCR).

References

Alexander AG (1985) The energy cane alternative. Elsevier, Amsterdam

Buckeridge MS, De Souza AP, Arundale RA, Anderson-Teixeira KJ, DeLucia E (2012) Ethanol from sugarcane in Brazil: a 'midway' strategy for increasing ethanol production while maximizing environmental benefits. GCB Bioenergy 4:119–126

Cançado JED, Saldiva PHN, Pereira LAA, Lara LBLS, Artaxo P, Martinelli LA, Arbex MA, Zanobett A, Braga ALF (2006) The impact of sugar cane–burning emissions on the respiratory system of children and the elderly. Environ Health Perspect 114: 725–729

Cantarella H, Cerri CEP, Carvalho JLN, Magalhaes PSG (2013) Editorial: How much sugarcane trash should be left on the soil. Scientia Agricola 70:1–2

Carmo JB, Filoso S, Zotelli LC, Sousa Neto E, Pitombo LM, Duarte Neto PJ, Vargas VP, Andrade CA, Gava GJC, Rossetto RR, Cantarella H, Elias Neto A, Martinelli LA (2013) Infield greenhouse gas emissions from sugarcane soils in Brazil: effects from synthetic and organic fertilizer application and crop trash accumulation. Glob Change Biol Bioenergy 5:267–280

Carvalho JLN, Otto R, Franco HCJ, Trivelin PCO (2013) Input of sugarcane post-harvest residues into the soil. Sci Agric 70:336–344

Carvalho-Netto OV, Bressiani JA, Soriano HL, Fiori CS, Santos JM, Barbosa GVS, Xavier MA, Landell MGA, Pereira GAG (2014) The potential of the energy cane as the main biomass crop for the cellulosic industry. Chem Biol Technol Agric 1(1):20. doi:10.1186/s40538-014-0020-2

Cerri CEP, Galdos MV, Carvalho JLN, Feigl BJ, Cerri CC (2013) Quantifying soil carbon stocks and greenhouse gas fluxes in the sugarcane agrosystem: point of view. Sci Agric 70:361–368

Davis SC, Parton WJ, Grosso SJ, Keough C, Marx E, Adler PR, DeLucia EH (2012) Impact of second-generation biofuel agriculture on greenhouse-gas emissions in the corn-growing regions of the US. Front Ecol Environ 10(2):69–74. doi:10.1890/110003

Dias MOS, Junqueira TL, Cavalett O, Pavanello LG, Cunha MP, Jesus CDF, Maciel R, Bonomi A (2013) Biorefineries for the production of first and second generation ethanol and electricity from sugarcane. Appl Energy 109: 72–78

Dinardo-Miranda LL, Fracasso JV (2013) Sugarcane straw and the populations of pests and nematodes. Sci Agric 70:369–374

Don A, Osborne B, Hastings A, Skiba U, Carter MS, Drewer J, Flessa H, Freibauer A, Hyvönen N, Jones MB, Lanigan GJ, Mander Ü, Monti A, Djomo SN, Valentine J, Walter K, Zegada-Lizarazu W, Zenone T (2012) Land-use change to bioenergy production in Europe: implications for the greenhouse gas balance and soil carbon. Glob Change Biol Bioenergy 4(4):372–391

Dondini M, Groenigen KJV, Galdo ID, Jones MB (2009) Carbon sequestration under Miscanthus: a study of 13C distribution in soil aggregates. Glob Change Biol Bioenergy 1:321–330

Duval BD, Anderson-Teixeira KJ, Davis SC, Keogh C, Long SP, Parton WJ, DeLucia EH (2013) Predicting greenhouse gas emissions and soil carbon from changing pasture to an energy crop. PLoS One 8(8). doi:10.1371/journal.pone.0072019

Figueiredo EB, Jr LS (2011) Greenhouse gas balance due to the conversion of sugarcane areas from burned to green harvest in Brazil. Agr Ecosyst Environ 141:77–85

Figueiredo EB, Panosso AR, Reicosky D, Jr LS (2014) Short-term CO2-C emissions from soil prior to sugarcane (Saccharum spp.) replanting in southern Brazi. Glob Change Biol Bioenergy 7(2):316–327. doi:10.1111/gcbb.12151

Franco HCJ, Pimenta MTB, Carvalho JLN, Magalhaes PSG, Rossell CEV, Braunbeck OA, Vitti AC, Kolln OT, Neto JR (2013) Assessment of sugarcane trash for agronomic and energy purposes in Brazil. Sci Agric 70:305–312

Galdos MV, Cerri CC, Cerri CEP (2009a) Soil carbon stocks under burned and unburned sugarcane in Brazil. Geoderma 153:347–352

Galdos MV, Cerri CC, Cerri CEP, Paustian K, Antwerpen R (2009b) Simulation of soil carbon dynamics under sugarcane with the CENTURY model. Soil Sci Soc Am J 73:802–811

Galdos MV, Cerri CC, Cerri CEP, Paustian K, Antwerpen R (2010) Simulation of sugarcane residue decomposition and aboveground growth. Plant Soil 326:243–259

Galdos MV, Cavalett O, Seabra JEA, Nogueira LAH, Bonomi A (2013) Trends in global warming and human health impacts related to Brazilian sugarcane ethanol production considering black carbon emissions. Appl Energy 104:576–582

Graham MH, Haynes RJ, Meyer JH (2002a) Changes in soil chemistry and aggregate stability induced by fertilizer applications, burning and trash retention on a long-term sugarcane experiment in South Africa. Eur J Soil Sci 53:589–598

Graham MH, Haynes RJ, Meyer JH (2002b) Soil organic matter content and quality: effects of fertilizer applications, burning and trash retention on a long-term sugarcane experiment in South Africa. Soil Biol Biochem 34:93–102

Grosso SJ, Parton WJ, Keough CA, Reyes-Fox M (2011) Special features of the DayCent modeling package. In: Ahuja LR, Ma L (eds) Methods of introducing system models into agricultural research. American Society of Agronomy, Madison

Hassuani SJ, Leal MRLV, Macedo IC (2005) Biomass power generation: sugar cane bagasse and trash, 1st edn. PNUD—Programa das Nações Unidas para o Desenvolvimento e CTC—Centro de Tecnologia Canavieira, Piracicaba

Hastings A, Clifton-Brown J, Wattenbach M, Mitchell CP, Smith P (2009) The Future energy potential of miscanthus in Europe. GCB Bioenergy 1:180–196

Horta Nogueira LA, Leal RLV (2012) Sugarcane bioethanol and bioelectricity. In: Poppe MK, Cortez LAB (eds) Sustainability of sugarcane bioenergy. CGEE—Centro de Gestão e Estudos e Estratégicos, Brasilia, pp 113–150

Humpenöder F, Schaldach R, Cikovani Y, Schebek L (2013) Effects of land-use change on the carbon balance of 1st generation biofuels: an analysis for the European Union combining spatial modeling and LCA. Biomass Bioenergy 56:166–178

Lal R (2008) Carbon sequestration. Philos Trans R Soc B Biol Sci 363:815–830

Lal R (2013) Soil carbon management and climate change. Cardiol Manage 4:439–462

Leal RLV, Galdos MV, Scarpare FV, Seabra JEA, Walter A, Oliveira COF (2013) Sugarcane straw availability, quality, recovery and energy use: a literature review. Biomass Bioenergy 53:11–19

Lee JA, Gill TE (2015) Multiple causes of wind erosion in the Dust Bowl. Aeolian Res 19:15–36

Li R, Merchant JW (2013) Modeling vulnerability of groundwater to pollution under future scenarios of climate change and biofuels-related land use change: a case study in North Dakota, USA. Sci Total Environ 447:32–45

Luca EF, Feller C, Cerri CC, arthes BB, Chaplot V, Campos DC, Manechini C (2008) Evaluation of physical properties and soil carbon and nitrogen stocks as affected by burning or green trash management of sugarcane. Rev Bras Ci Solo 32:789–800

Lugato E, Jones A (2014) Modelling soil organic carbon changes under different maize cropping scenarios for cellulosic ethanol in Europe. Bioenergy Res 8:537. doi:10.1007/s12155-014-9529-2

Marin FR, Thorburn PJ, Costa LG, Otto R (2014) Simulating long-term effects of trash management on sugarcane yield for Brazilian croppint systems. Sugar Tech 16:164–173

Masters MD, Black CK, Kantola IB, Woli KP, Voigt T, David MB, DeLucia EH (2016) Soil nutrient removal by four potential bioenergy crops: *Zea mays*, *Panicum virgatum*, *Miscanthus x giganteus*, and prairie. Agric Ecosyst Environ 216:51–60

Matsuoka S, Kennedy AJ, Gustavo E, Santos D, Tomazela AL, Rubio LCS (2014) Energy Cane: its concept, development, characteristics, and prospects. Adv Bot 2014

Miyake S, Renouf M, Peterson A, McAlpine C, Smith C (2012) Land-use and environmental pressures resulting from current and future bioenergy crop expansion: A review. J Rural Stud 28: 650-658

Moraes BS, Zaiat M, Bonomi A (2015) Anaerobic digestion of vinasse from sugarcane ethanol production in Brazil: challenges and perspectives. Renew Sustain Energy Rev 44:888–903

Muth DJ, McCorkle DS, Koch JB, Bryden KM (2012) Modeling sustainable agricultural residue removal at the subfield scale. Agron J 104(4):970

Odell RT (1982) The Morrow plots: a century of learning. Bulletin, Agricultural Experiment Station, University of Illinois at Urbana-Champaign

Oliveira BG, Carvalho JLN, Cerri CEP, Cerri CC, Feigl BJ (2013) Soil greenhouse gas fluxes from vinasse application in Brazilian sugarcane areas. Geoderma 200–201:77–84

Otto R, Franco HCJ, Faroni CE, Vitti AC, Trivelin PCO (2009) Fitomassa de raízes e da parte aérea da cana-de-açúcar relacionada à adubação nitrogenada de plantio. Pesq Agrop Brasileira 44:398–405

Paredes DS, Lessa ACR, Selenobaldo ACS, Boddey RM, Urquiaga S, Alves BJR (2014) Nitrous oxide emission and ammonia volatilization induced by vinasse and N fertilizer application in a sugarcane crop at Rio de Janeiro, Brazil. Nutr Cycl Agroecosyst 98:41–55

Paustian K, Collins HP, Paul EA (1997) Management controls on soil carbon. In Paul EA, Paustian K, Elliott ET, Cole CV (Ed.) Soil organic matter in temperate agroecosystems: long-term experiments in North America. Boca Raton: CRC press. p.15-49

Pinheiro EFM, Lima E, Ceddia MB, Urquiaga S, Alves BJR, Boddey RM (2010) Impact of pre-harvest burning versus trash conservation on soil carbon and nitrogen stocks on a sugarcane plantation in the Brazilian Atlantic forest region. Plant Soil 333:71–80

Ratnavathi CV, Kalyana Chakravarthy S, Komala VV, Chavan UD, Patil JV (2011) Sweet Sorgum as feedstock for biofuel production: a review. Sugar Tech 13(4):399–407

Resende AS, Xavier RP, Oliveira OC, Urquiaga S, Alves BJR, Boddey RM (2006) Long-term effects of pre-harvest burning and nitrogen and vinasse applications on yield of sugar cane and soil carbon and nitrogen stocks on a plantation in Pernambuco, NE Brazil. Plant Soil 281:339–351

Robertson F (2003) Sugarcane trash management: consequences for soil carbon and nitrogen – Final Report to the CRC for Sustainable Sugar Production of the project Nutrient Cycling in Relation to Trash Management. Townville: CRC for Sustainable Sugar Production. 39p.

Robertson FA, Thorburn PJ (2007) Management of sugarcane harvest residues: consequences for soil carbon and nitrogen. Aust J Soil Res 45:13–23

Rossetto R, Dias FLF, Landell MGA, Cantarella H, Tavares S, Vitti AC, Perecin D (2010) N and K fertilisation of sugarcane ratoons harvested without burning. Proc Int Soc Sugar Cane Technol 27:1–8

Sheehan J, den AA, Paustian K, illian KK, Brenner J, Walsh M, Nelson R (2003) Energy and environmental aspects of using corn stover for fuel ethanol. J Ind Ecol 7(3–4):117–146

Silva GRV, Souza ZMD, Martins Filho MV, Barbosa RS, Souza GS (2012) Soil, water and nutrient losses by interrill erosion from green cane cultivation. Rev Bras Ci Solo 36:963–970

Silver WL, Neff J, McGroddy M, Veldkamp Ed, Keller M, Cosme R (2000) Effects of Soil Texture on Belowground Carbon and Nutrient Storage in a Lowland Amazonian Forest Ecosystem. Ecosystems 3: 193-209

Siqueira Neto M, Galdos MV, Feigl BJ, Cerri CEP, Cerri CC (2015) Direct N_2O emission factors for synthetic n-fertilizer and organic residues applied on sugarcane for bioethanol production in central-southern brazil. Glob Change Biol Bioenergy 8(2):269–280

Smith JU, Gottschalk P, Bellarby J, Chapman S, Lilly A, Towers W, Bell J, Coleman K, Nayak DR, Richards MI, Hillier J, Flynn HC, Wattenbach M, Aitkenhead M, Yeluripurti JB, Farmer J, Milne R, Thomson A, Evans C, Whitmore AP, Falloon P, Smith P (2010) Estimating changes in national soil carbon stocks using ECOSSE—a new model that includes upland organic soils. Part I. Model description and uncertainty in national scale simulations of Scotland. Climate Res 45:179–192

Sordi G, Mane c (2013) Utilization of trash: a view from the agronomic and industrial perspective. Sci Agric 70(3–4) Opinion

Souza AP, Arundale RA, Dohleman FG, Long SP, Buckeridge MS (2013) Will the exceptional productivity of Miscanthus x giganteus increase further under rising atmospheric CO_2? Agric For Meteorol 171/172:82–92

Signor D, Pissioni LLM, Cerri CEP (2014) Emissões de gases de efeito estufa pela deposição de palha de cana-de-açúcar sobre o solo. Bragantia. doi:10.1590/brag.2014.019

Thorburn PJ, Meier EA, Collins K, Robertson FA (2012) Changes in soil carbon sequestration, fractionation and soil fertility in response to sugarcane residue retention are site-specific. Soil Tillage Res 120:99–111

Trivelin PCO, Franco HCJ, Otto R, Ferreira DA, Vitti AC, Fortes C, Faroni CE, Oliveira ECA, Cantarella H (2013) Impact of sugarcane trash on fertilizer requirements for São Paulo, Brazil. Sci Agric 70:345–352

UNICA (2015) http://www.unica.com.br/noticia/6551584920310621254/sao-paulo-fecha-safra-2013-por-cento2F2014-com-colheita-mecanizada-em-83-por-cento-dos-canaviais/. Accessed 23 Mar 2015

Vargas VP, Cantarella H, Martins AA, Soares JR, Carmo JB, Andrade CA (2014) Sugarcane crop residue increases N2O and CO2 emissions under high soil moisture conditions. Sugar Tech 16:174–179

Viator RP, Johnson RM, Grimm CC, Richard EP Jr (2006) Allelophatic, autotoxic, and hormetic effects of postharvest sugarcane residue. Agron J 98:1526–1531

Vitti AC, Franco HCJ, Trivelin PCO, Ferreira DA, Otto R, Fortes C, Faroni CE (2011) Nitrogênio proveniente da adubação nitrogenada e de resíduos culturais na nutrição da cana-planta. Pesq Agrop Brasileira 46:287–293

Wisniewski C, Holtz GP (1997) Decomposição da palhada e liberação de nitrogênio e fosforo numa rotação aveia-soja sob plantio direto. Pesq Agrop Brasileira 32:1191–1197

Zatta A, Clifton-Brown J, Robson P, Hastings A, Monti A (2013) Land use change from C3 grassland to C4 Miscanthus: effects on soil carbon content and estimated mitigation benefit after six years. Glob Change Biol Bioenergy 6:360–370

Zeri M, Anderson-Teixeira K, Hickman G, Masters M, DeLucia E, Bernacchi CJ (2011) Carbon exchange by establishing biofuel crops in Central Illinois. Agr Ecosyst Environ 144:319–329

Zimmerman AR, Gao B, Ahn MY (2011) Positive and negative carbon mineralization priming effects among a variety of biochar-amended soils. Soil Biol Biochem 43: 1169–1179

Chapter 11
Policy and Diplomacy in the Production of Second Generation Ethanol in Brazil: International Relations with the EU, the USA and Africa

Nicola Favretto, Lindsay C. Stringer, Marcos S. Buckeridge, and Stavros Afionis

Abstract Concerns over the global energy crisis and climate change are increasingly mediated through the deployment of renewable energy sources, such as biofuels. Second generation (2G) biofuels derived from non-edible feedstocks offer more sustainable and higher productive alternatives than first generation (1G) ones, which are more water and land intensive. New cooperation opportunities, particularly among Brazil, the USA and the EU, have been generated by the emerging 2G global market. Future expansion of 2G production and consumption in the African continent is envisaged to offer a range of benefits, including decreased oil dependency and generation of economic activities which can foster rural development. This chapter outlines the current state of affairs with respect to 2G biofuels in Brazil, including discussion on the future prospects of introducing 2G production in Africa. Opportunities and challenges of 2G cooperation between Brazil, the USA and the EU are discussed. Secondary data from government documents, published reports and scientific studies were integrated with semi-structured interviews and personal communications with Brazilian line ministries and EU officials. Data show a great potential for the expansion of Brazil's 1G ethanol industry in Africa, but 2G development is not identified as a foreseeable option in the forthcoming years. This is due to the high production costs, lack of advanced technology and the unattractive business environment. The creation of a 1G consumer market and establishment of safe legal and policy frameworks are needed in order to foster 2G investments.

N. Favretto (✉) • L.C. Stringer • S. Afionis
School of Earth and Environment, Sustainability Research Institute,
University of Leeds, Leeds, UK
e-mail: n.favretto@leeds.ac.uk; S.Afionis@leeds.ac.uk

M.S. Buckeridge
Laboratory of Plant Physiological Ecology (LAFIECO), Department of Botany,
Institute of Biosciences, University of São Paulo, Sao Paulo, Brazil

M.S. Buckeridge, A.P. De Souza (eds.), *Advances of Basic Science for Second Generation Bioethanol from Sugarcane*, DOI 10.1007/978-3-319-49826-3_11

As observed in terms of international cooperation, Brazil's potential to foster 2G technologies in Africa is compromised by a disconnect between science production, industrial and policy developments. Cooperation of Brazil with the EU is focused on fundamental sciences, while linkages with the USA are limited to private sector and market development. Increased policy coordination towards the transition from 1G to 2G, as well as stronger budget support, is key for enabling Brazil's future leadership in the international 2G arena.

Keywords Ethanol diplomacy • European Union • Africa • United States of America • Bioethanol • Sugarcane

11.1 Introduction

Between 2010 and 2040 the world's energy consumption is estimated to increase at a rate of 56%, from 524 quadrillion British thermal units (Btu) in 2010 to 820 quadrillion Btu in 2040 (EIA 2013). Driven by rapid population growth and economic expansion, more than 85% of this unprecedented growth will occur among developing countries, with petroleum, natural gas and coal dominating the energy mix (EIA 2013). Such high reliance on fossil fuels has raised concerns about energy security across the globe, particularly in the context of fluctuating global oil prices (Hamilton 2009), scarcity of available petroleum reserves (Sorrell et al. 2010) and political instability in the regions in which these reserves are located (Luft and Korin 2009). Concerns related to climate change are also pertinent, as combustion of fossil fuels remains the largest contributing factor to the release of greenhouse gases (GHG) into the atmosphere (IPCC 2014).

To address the aforementioned complications, governments worldwide have progressively introduced incentives to support the deployment of renewable energy sources, like *inter alia* biofuels (EIA 2013). First generation biofuels (1G) derived from edible agricultural products have been promoted as an option to enhance access to energy, lessen dependence on expensive imported oil, as well as reduce GHG emissions (Janssen and Rutz 2012; Goldemberg 2008). As a result, global 1G biofuel supply chains have been developed, particularly in the transport sector, where biofuels output, adjusted for energy content, accounted for 3.5% of global oil demand for road transport in 2013 (IEA 2014).

However, 1G biofuels are mainly produced from food crops, such as grains, sugarcane and vegetable oils. Their capability to contribute to long-term fossil fuel substitution has been questioned, as a range of debates on their socio-economic and environmental sustainability have been raised, particularly with regard to their impacts on food security and land-use (Nonhebel 2012; Cotula et al. 2008). More advanced fuels can be derived from lignocellulosic biomass (e.g. timber and waste products from forestry, agriculture, industry or households), specific non-edible energy crops (e.g. sugarcane bagasse, switchgrass, miscanthus and willow), biomass-to-liquids (BtL)-diesel and bio-synthetic gas (bio-SG) (IEA 2011). These fuels, collectively referred to as second generation biofuels (2G), are thought to offer more sustainable alternatives than 1G as they are more energy efficient, less water and land intensive (Hill 2007), and are expected to achieve greater GHG emissions reductions (Havlík et al. 2011).

From a global perspective, according to Köhler et al. (2014), a total of 69 2G pilot plants were established worldwide as of 2014, of which 35 were based in Europe, 21 in the USA, 6 in Brazil, 5 in Canada and 2 in Asia. Even though some of these plants have the technical capabilities to produce small quantities of cellulosic 2G, the industry has not yet reached commercial deployment and production technology remains still at a development stage. Research and development (R&D) efforts have not yet produced solid results in the developed world and estimates about the timing for 2G's commercial scaling-up remain uncertain. Varied estimates of possible growth paths are provided in the literature (Khanna 2008). According to interviews with Petrobras Biofuel[1] and the Sugarcane Technology Research Center[2] in Brazil, commercial scaling-up should be achieved by the late 2010s.

As a consequence of supportive legislation and adequate funding spearheaded by the USA and, to a lesser extent, the European Union (EU) and Brazil, the future role of 2G global production and consumption is expected to become more prominent. Starting with the USA, its 2007 Renewable Fuel Standard sets a consumption mandate ranging between 0.1 billion gallons of lignocellulosic biomass in 2010 and 16 billion gallons in 2022 (Government of the US 2007). These ambitious policy goals resulted in growing R&D investments, which have led to the implementation of a number of pilot projects in the 2G sector (USDOE 2008). Yet, actual production in the USA remains far below policy targets (i.e. no 2G was used before 2012 and only 200,000 gallons were commercialised in 2012). As a result, in 2013 the 2G consumption mandate was waived by the Environmental Protection Agency (EPA) from the initial 1 billion gallons to 0.014 billion gallons (Thompson and Meyer 2013). Annual percentage standards for 2014, 2015 and 2016 were set to 0.033, 0.106 and 0.206 billion gallons, respectively (EPA 2015).

Turning to the EU, since the EU Biofuels Directive was approved in 2003, a range of policy instruments have been adopted with the aim to promote biofuel production. These include subsidy schemes, tax exemptions, funding for R&D and Common Agricultural Policy supporting mechanisms (Afionis and Stringer 2012). The EU's more recent 2009 Renewable Energy Directive (RED) sets a mandatory target for energy from renewable fuels in transport of 10% by 2020 (European Commission 2009) and envisages the creation of a 1G market as a starting point for the development of a 2G one (Baka and Roland-Holst 2009). Benefits to the latter are provided in the RED, where contributions of 2G are credited with a multiplier of 2 towards the 10% target (European Commission 2009). Until 2015, EU actors had been negotiating amendments to RED, focusing in particular on whether the 10% target should be differentiated to expressly include a sub-target for advanced

[1] Interview, Petrobras Biocombustível, Rio de Janeiro, March 2013.

[2] Interview, Sugarcane Technology Research Center (CTC), Piracicaba, March 2013.

biofuels. Intra-EU negotiations to this effect commenced in 2010, and were only concluded in April 2015 with a political compromise to set a 7% cap on 1G food-crop biofuels and encourage Member States to individually put in place indicative targets for 2G and 3G (advanced) biofuels. In October 2014, the EU agreed on a new framework for climate and energy, which included EU-wide targets and policy objectives for the period between 2020 and 2030. One of these targets calls for at least a 27% share of renewable energy consumption (European Commission 2015). As of April 2016, the EU is in the process of reviewing its bioenergy sustainability policy in order to propose a new RED for the period after 2020.

Brazil is widely recognised as a global leader in 1G production, being the second ethanol producer worldwide after the USA and the leading exporter, accounting for more than 40% of global trade (Walter and Segerstedt 2012). While Brazil's 1G leadership dates back to the 1970s, ethanol production and consumption were principally boosted following the introduction of flex-fuel car engines into the market in 2003, which could run on 100% gasoline, 100% bioethanol or any proportion mix of the two without any effect on vehicle performance (Afionis et al. 2013).

Due to various political and economic factors, however, the Brazilian sugarcane sector is today undergoing a period of deep crisis and even Brazil has experienced a dampening of enthusiasm for biofuels (Afionis et al. 2016). Internationally, concerns about biofuels sustainability, coupled with the turbulence generated by the global and national economic crises, halted Brazil's export-oriented ethanol production expansion plans. Domestically, poor plantation management and planning decisions effectively crippled ethanol production in Brazil during the early 2010s, causing market uncertainty, driving ethanol prices up and forcing Brazil to import ethanol from the USA in 2011 (Afionis et al. 2013).

Developing 2G biofuels has therefore been envisaged as a promising remedial solution, given their higher productivity and sustainability credentials. Indeed, almost all crops have more sugar potentially available in the cell walls of their stems and leaves than in their grains (SUNLIBB 2014). While the current inefficiency of conversion from biomass to biofuels makes 2G production economically uncompetitive, productivity gains related to the learning-curve effect are expected to eventually have a tapering effect on cost. According to Brazilian industry stakeholders, 2G biofuels could potentially be as much as 20% cheaper compared to their 1G counterparts in a decades' time.[3] While no formal legislation has been promulgated up to date, substantial financial and technological resources have been devoted into developing Brazil's 2G capacity. For instance, Brazilian public actors, like BNDES[4] and FINEP,[5] have cumulatively contributed close to US$ 2 billion since 2011 in 2G innovation projects, up from just US$ 100 million that same year.[6] However, the investment made was not adequately coupled with the produc-

[3] Telephone interview, GranBio, a 2G commercial plant in Alagoas, Brazil (September 2014).

[4] Banco Nacional de Desenvolvimento Econômico e Social.

[5] Financiadora de Estudos e Projetos; an organisation under the Brazilian Ministry of Science & Technology.

[6] Interview, BNDES, Sao Paulo (September 2014).

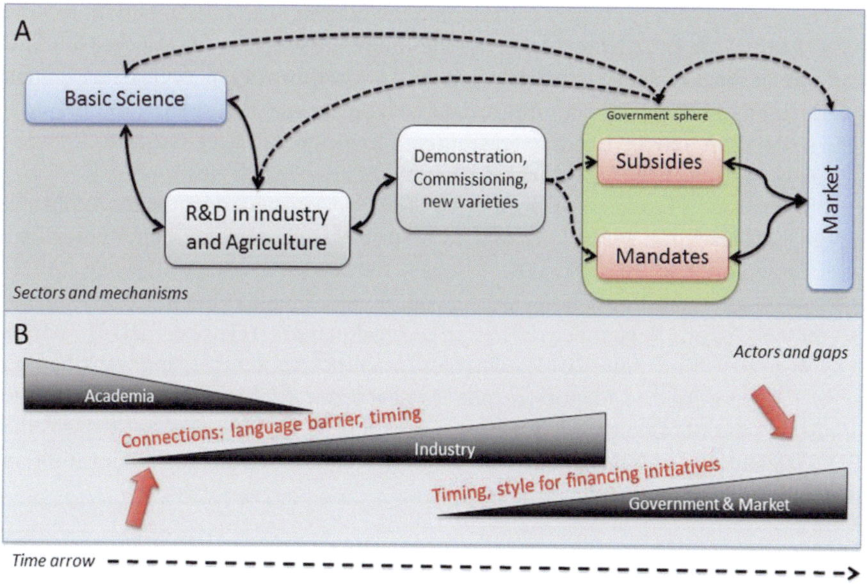

Fig. 11.1 Failures in the model for an innovation system for ethanol 2G production in Brazil. The natural sequence of events is shown in A, with *solid arrows* being the exchanges of information among actors. *Dashed arrows* are irreversible events. B shows the main problems related to the exchange of information between the main actors in the innovation game (academia, industry, government and market). The main reasons for system's failure are given in *red* and pointed with thick *red arrows*. When connections, timing and style of financing initiatives are well adjusted, the innovation system will work more efficiently and more quickly. When one of them does not work efficiently, the whole system fails

tion of basic and applied sciences related to 2G technologies, leading to a disconnection between the production of knowledge and technology. As a result, 2G technologies currently used in Brazil are mostly based on science produced outside of the country (Fig. 11.1).

The emerging 2G global market has also opened up new cooperation opportunities for major actors like Brazil, the USA and the EU, which have been progressively engaging in collaborative efforts aimed at overcoming the current production barriers. International cooperation is thought to offer not only attractive opportunities for more sustainable and profitable biofuel production, but also the prospect to establish biofuel markets elsewhere, for instance, in Africa. In addition to public-led initiatives, private sector companies from the aforementioned actors are actively investing in the Brazilian 2G sector, drawn by the country's feedstock availability and production capacity. Brazil's engagement in the 2G sector is expected to benefit the country by serving the twofold purpose of stabilising its internal biofuels market and reviving its ethanol diplomacy undertakings abroad (Afionis et al. 2016). However, difficulties in linking the scientific research carried out by Brazilian institutions to the country's international cooperation undertakings have resulted in limited capacity to foster 1G and 2G technologies in Africa to date.

The purpose of this chapter is to outline the current state of affairs with respect to 2G biofuels in Brazil, and to discuss the future prospects of introducing 2G biofuel production in African countries. Section 11.2 outlines 2G biofuels development in Brazil and reviews the opportunities and challenges of 2G cooperation between Brazil, the USA and the EU. The feasibility of expanding the 2G market to Africa is discussed in Sect. 11.3, while Sect. 11.4 summarises our findings.

Our analysis is based on secondary data from government documents, published reports and scientific studies, as well as interviews and personal communications with Brazilian and EU officials. Primary data were gathered through thirty-three semi-structured interviews carried out across Brazil (March 2013, August–September 2014, September 2015), EU headquarters (October 2013), Africa (October 2013 and July 2014), whereby a purposive sample and snowballing approach were used to identify respondents. Interviewees included Brazilian line ministries, the EU Delegation to Brazil, the European Commission in Brussels, the EU Delegations to Kenya and Mozambique, and the Brazilian Delegation to Brussels.

11.2 Potential for 2G Cooperation Between the EU, the USA and Brazil

During the 2000s, under the government of President Lula, Brazil pursued a range of coalitions with powerful 1G biofuels producers and consumers, such as the EU and the USA, with the aim to transform bioethanol into a globally traded commodity. Fundamentally, the creation of a wide 1G international market of both producers and consumers is required for commoditisation to take place: '*If you wish to make biofuel a commodity, more countries have to produce it*'.[7] In addition, Brazil is making an effort to stimulate 2G ethanol production, with two plants having already started production (see below). For Brazil, 2G is vital in order to meet domestic ethanol demand for 2020.[8] To create synergies, Brazil has engaged with the EU and the USA in setting up various projects for 1G and 2G ethanol production, with some of these initiatives even involving developing countries in Africa, Central America and the Caribbean.

Starting with the EU, its increasing preoccupation with climate change and energy security might induce an observer to assume—at first reading at least—that the EU and Brazil would be natural partners in promoting the use of sustainable biofuels on the global scene. Yet, our interviews point to the opposite conclusion. In particular, a difficult interaction with the European institutions has been noted by Brazil's policymakers, who claim that the counterpart is not willing to undertake concrete actions in the sector: '*They [the EU] say that they are maintaining an*

[7] Interview, Ministry of External Relations, Brasilia, March 2013.

[8] Telephone interview, GranBio, a 2G commercial plant in Alagoas, Brazil (September 2014).

energy dialogue, but it's nothing more than just talk'.[9] Another interviewee noted that with the EU there is a lot of talking, lots of repetitive meetings and, hence, lots of 'wasted' time.[10]

Over time, a number of emerging issues have greatly polarised biofuel relations between the two partners. A first issue relates to the EU-wide standards and certification criteria that have been promulgated by the RED with the intention to foster a more sustainable biofuels industry.[11] Even though the EU stresses the need to draw an explicit link between consumption of biofuels and their sustainable production, Brazil's policymakers perceive the EU's criteria as an instrument used to create barriers to international trade: '*There is a lobby against biofuels*',[12] '*The discussion about the sustainability criteria is a façade, an excuse to set up some barriers'*.[13] A second issue relates to market distortions caused by the EU's unfavourable trade regime, with import tariffs and tax credits being accused of offsetting the comparative advantage of leading producers like Brazil and constraining the commercialisation of their ethanol in the EU (Afionis and Stringer 2014).

This tense situation has had a knock-on effect in terms of fostering EU–Brazil partnerships in the 2G sector, which according to BNDES would represent an opportunity for Brazil to promote industrial scaling-up and maintain the country's technological leadership in the sugarcane industry, while at the same time supplying global markets with sustainable biofuels.[14] To date, cooperation is mostly concentrated in the areas of science, technology and innovation, with a number of joint projects having been funded through partnerships with individual Member States like the United Kingdom or Germany, or through the EU's Framework Programmes for Research and Technological Development (FPs). SUNLIBB[15]–CeProBio[16] was an example of such a project, which aimed *inter alia* at combining European and Brazilian research strengths in order to open the way for cost-competitive 2G biofuels production.

Scientific collaboration between Brazil and the EU on bioenergy can be considered as successful. SUNLIBB–CeProBio and other major scientific projects (e.g. INCT-Bioetanol) have been actively producing scientific data and have enriched the knowledge about bioenergy feedstocks on both sides of the Atlantic. The connections between Brazilian, UK, Dutch, French and German scientists in this area have generated a range of publications, facilitated, among others, by frequent joint meet-

[9] Interview, Ministry of Mining and Energy, Brasilia, March 2013.

[10] Interview, Ministry of Agriculture, Brasilia, September 2014.

[11] Besides greenhouse gas savings (currently 35 %, rising to 50 % in 2017), the EU's sustainability criteria stipulate that biofuel feedstock is not to be derived from primary forests, lands with high biodiversity value, protected territories and carbon-rich areas.

[12] Interview, Ministry of External Relations, Brasilia, March 2013.

[13] Interview, Institute of Electrotechnics and Energy, University of Sao Paulo, February 2013.

[14] Interview, BNDES, Sao Paulo, March 2013.

[15] Sustainable Liquid Biofuels from Biomass Biorefining (SUNLIBB). For further information, see sunlibb.eu.

[16] Centre of Biological and Industrial Process for Biofuels (CeProBio).

ings and a constant exchange of research staff. Throughout the duration of the projects, fundamental scientific advances have been achieved, leading to an increased understanding of 2G processes. Structure and composition of biomass feedstock for bioenergy in Brazil and Europe were studied in depth and are now better understood, while dozens of genes and enzymes were discovered and fully characterised. Plant growth and crop sustainability were also investigated by joint research groups in both regions.

In 2013, however, the planned publication for 2014 of a joint call on 2G biofuels that was to be funded under the remit of Horizon 2020 was delayed by the Brazilian side, as funds were reallocated to the Science without Borders initiative.[17] The latter is a Brazilian government scholarship programme, which aims to send in excess of 100,000 undergraduate and postgraduate Brazilian students at top universities around the world. Despite the setback, the call was eventually launched in October 2015.

Unlike the slow bilateral progress, European private sector 2G investments and collaborations in Brazil are substantial, even though they pale in comparison to those of their US counterparts. For example, the design and equipment of the Brazilian 2G plant in Alagoas was supplied by the Mossi Ghisolfi Group, which owns Europe's largest commercial-scale plant for the production of bioethanol from non-food biomass sources in Crescentino, Italy.[18] Furthermore, Shell in 2011 launched a joint venture with Cosan, Brazil's largest ethanol exporter, with one of the outcomes being the aforementioned 2G Raisen plant in Piracicaba. In December 2014, Abengoa and BNDES announced they would co-finance the construction of a 2G ethanol plant in Pirassununga, Sao Paulo, expected to be in full operation by 2017.[19]

However, a lack of coordination and communication between the industrial initiatives and the science being developed by Brazilian and European scientists has been observed. Such disconnect between science and industry is explained by the limited connectivity between the political and diplomatic parts of each region. This may be so because the latter are the forces that should be adjusting the timing of when and how the scientific knowledge would be directly applied in industry to generate maximum profit for both sides.

When it comes to relations with the USA, the overall picture is totally different, with Brazilian policymakers praising the ease with which they can cooperate with their American counterparts. One interviewee noted that the USA has clear goals and tools to implement actions,[20] while another argued that it is easier to work with them given their similar views on sustainable biofuels: '*We do not have to fight about the concept, as is the case with the EU*'.[21] Indicative of the positive bilateral relationship is the decision by the US government to satisfy a long-standing demand and allow the US$0.54 per-gallon surcharge on imported ethanol to expire at the end

---

[17] Interview, EU Delegation to Brazil, Brasilia, September 2014.

[18] Interview, BNDES, Sao Paulo, September 2014.

[19] Interview, CTBE, Campinas, Sao Paulo, September 2015.

[20] Interview, BNDES, Sao Paulo, September 2014.

[21] Interview, Ministry of External Relations, Brasilia, September 2014.

of 2011, a measure initially put in place to shelter ethanol distilled from corn in the USA from ethanol distilled from sugarcane in Brazil.[22]

In terms of bilateral relations, Brazil and the USA signed a Memorandum of Understanding (MoU) in March 2007 to cooperate on biofuels, which was extended in 2011 to include aviation biofuels. Since then, they have been engaged in various R&D cooperation undertakings, focusing, for example, on maximising the efficiency of flex-fuel vehicle engines or on facilitating the commercial-scale development and deployment of aviation biofuels. Under the MoU, a subset of the activities is directed towards the stimulation of private sector biofuels investment in third countries. While initially the focus was solely on Central American and Caribbean countries, it was later expanded to include African countries, like Senegal and Guinea-Bissau (Dalgaard 2012). Technical assistance aims to strengthen policy frameworks, implement blending laws, and develop domestic biofuels production capabilities.[23]

In terms of private sector investments, US companies have been more active in Brazil than EU ones.[24] To name a few examples, Solazyme, a San Frasisco-based bioproducts company, has a facility in Sao Paulo State and was planning to use modified algae to produce oil from sugarcane to be used in *inter alia* bio jet fuel, speciality chemicals and food supplements.[25] Amyris is another US company that has production facilities in Sao Paulo, investing among others in aviation biofuels from sugarcane. Amyris has partnered with Brazilian airline GOL, and in July 2014 a demonstration flight using farnesane was performed from Orlando, Florida, to Sao Paulo, Brazil.[26] Finally, Boeing has signed a MoU with Brazilian aerospace conglomerate Embraer, to invest in a joint biofuels research centre in Sao Paulo State for developing sustainable aviation biofuels (Boeing 2014).

However, unlike cooperation with the EU, no joint fundamental sciences projects were established. US initiatives were always a ready-made strategy, bringing American technology to exploit the Brazilian market. In other words, there were no partnerships to support win-win cooperation in terms of common goals to achieve major objectives in the sustainable production of biofuels. Thus, the relationship with the USA can be characterised as even more chaotic and unilaterally focused on profitable opportunities, rather than anchored in cooperation between countries that would lead to a better world. In this sense, although some connections have been seen between Brazil and the EU, US–Brazil relations could be considered rather disparate.

Overall, expectations were that if these aforementioned 2G biofuel innovative investments proved successful, and as the global economic crisis recedes, more and more companies from across the world would be drawn to Brazil to seek joint ven-

[22] Interview, Institute of Energy and Environment, University of Sao Paulo, August 2014.

[23] Interview, US Embassy in Brazil, September 2014.

[24] Interview, Ministry of Agriculture, Brasilia, September 2014.

[25] Interview, US Embassy in Brazil, September 2014.

[26] Interview, US Embassy in Brazil, September 2014.

tures and invest.[27] However, due to Brazil's unstable economic and political situation, all such expectations need to be put on hold for the time being. In May 2016, Brazilian President Dilma Rousseff was suspended from office pending an impeachment trial over accusations that she illegally manipulated government accounts. Consequently, even though Brazilian institutions, like BNDES, as well as Brazilian industrial stakeholders have been paying extra attention in ensuring the success of 2G undertakings, the crisis in Brazil has created uncertainty as to the continuity prospects of such strategies.

Three cautious notes are in order at this point. Development of 2G biofuels has the potential to help, but is not going to be the answer to all of Brazil's bioethanol problems. First, bringing back stability to its 1G sector is a prerequisite for successfully consolidating a 2G market and reviving anew its ethanol diplomacy overtures abroad. After all, the latter two presuppose a healthy sugarcane industry that is able to produce adequate supplies of feedstock. Second, Brazil's policy remains focused on the promotion of 1G rather than 2G. No legislation promoting 2G is currently in place at the national level, and according to Brazil's 10-Year National Energy Expansion Plan 2012–2021, the 61 billion litres of ethanol expected to be produced in 2021 will come entirely from 1G feedstocks (Government of Brazil 2013). The urgent need to elaborate a mandate and rules aimed at creating suitable business conditions for 2G is stressed by key producers: '*If the Brazilian government wants to lead this [2G development], we have to think about having a mandate to create the business conditions to compete with the most advanced foreign countries (like the US) on 2G*'.[28] Clearer rules are also needed in terms of how varied combinations of 1G and 2G will be accounted for: '*Certainly the final product sold in the market will be a mix of first and second generation fuels. Otherwise, if we isolate the plant, production would not be economically feasible. So, which certificate will accept this mix? Under what conditions?*'[29] Third, the problem with science-policy interfaces needs to be addressed so as to better incorporate the fundamental science produced during the last decade in Brazil into the various public and private sector undertakings that have been emerging there. Doing so will provide industry with a more solid grounding and fewer trial-and-error strategies in the production of 2G ethanol.

11.3 Expanding Brazil's 2G Production in Africa: An Ambitious Goal or Achievable Reality?

As noted in the preceding sections, 2G biofuels may play a major role in enabling Brazil to stabilise its internal ethanol market, as well as revitalise its ethanol diplomacy strategy. So far the latter has produced limited results due to a multiplicity of reasons, most important being lack of policy guidance on the part of the Rousseff

[27] Telephone interview, GranBio, a 2G commercial plant in Alagoas, Brazil (September 2014).

[28] Interview, Petrobras Biocombustível, Rio de Janeiro, March 2013.

[29] Interview, Petrobras Biocombustível, Rio de Janeiro, March 2013.

administration, domestic production shortages, as well as the increasingly hostile international environment for 1G biofuels. However, the advantages of 2G biofuels address most of these issues to some degree, thereby potentially allowing Brazil to strengthen its ethanol promotion efforts abroad in the future.

The greatest potential for the expansion of Brazil's ethanol production is found in the African continent due to the favourable climatic and soil conditions suitable to sugar cane plantation (Guardabassi and Goldemberg 2014), the perceived availability of agricultural land (Dalgaard 2012; Lynd and Woods 2011), and the shared cultural values and economic interests across Portuguese-speaking African counties (Cabral et al. 2013). The establishment of a global biofuel market is foreseen to deliver a range of benefits to African countries, including decreased dependency on oil and reduced foreign debt at the national level, as well as the generation of economic activities fostering rural development at the local level (Walter and Segerstedt 2012; Favretto et al. 2013).

Moved by these drivers, Brazil has actively engaged in cooperation with African countries towards their expansion of 1G production: '*It is our interest to sell our technology…but we believe that we can also help them [Africa] in South-South cooperation… We try to really help in the sense of leaving them our expertise*'.[30] Indicative of the importance Brazil places in such interactions is that ethanol diplomacy found its way into Brazil's pledge for the 2015 Paris climate summit. In particular, its Intended Nationally Determined Contribution states that: 'Brazil will undertake best efforts to enhance cooperation initiatives with other developing countries, particularly in the areas of… biofuels capacity-building and technology transfer' (UNFCCC 2015).

Two routes have been pursued so far, namely the establishment of direct South–South bilateral cooperation, as well as the initiation of North–South–South trilateral partnerships involving the EU or the USA as a Northern donor. In pursuing the bilateral path, technical cooperation agreements have been put in place with various countries, like Mozambique or Angola, with the aim of implementing joint projects led by Brazilian agencies and the private sector. Through the Getúlio Vargas Foundation (Fundação Getúlio Vargas—FGV), a number of feasibility studies have been conducted in the continent so as to identify optimal production sites, technologies and feedstocks (FGV 2009). The most promising countries identified by the FGV's studies include Senegal, Zambia and Mozambique.[31] Brazilian private sectors actors, such as Petrobrás, Guarani and Odebrecht, have already entered the African market, having invested in sugar mills and 1G ethanol plants in Mozambique and Angola.[32]

Turning to trilateral cooperation initiatives involving the EU, it had been decided during the third EU-Brazil Summit in Stockholm in 2009 to set up joint trilateral projects for bioenergy and ethanol production in Africa, with Kenya and Mozambique serving as pilot countries. However, the two projects eventually became embroiled

[30] Interview, Ministry of External Relations, Brasilia, March 2013.

[31] Interview, Ministry of Mining and Energy, Brasilia, March 2013.

[32] Interview, Ministry of External Relations, Brasilia, September 2014.

in the international sustainability controversy and were therefore never actually implemented.[33] As far as trilateral cooperation involving the USA is concerned, outcomes have been more substantial, with various African, Central American and Caribbean countries having received help through feasibility studies and technical assistance on regulatory framework development (Afionis and Stringer 2015).

Nevertheless, a range of economic, legal and regulatory constraints have posed significant barriers to 1G development. An obvious complication relates to the lack of capital, policies, infrastructure and skilled labour, as well as unclear land tenure regimes. Another commonly noted problem is the lack of legal and regulatory frameworks, which can cause market and investor uncertainty. Unstable economic and political conditions also hinder investments, as Brazilian funding institutions, like the BNDES, face difficulties in covering the requirements and mitigation costs linked to issues such as currency fluctuations and civil wars.[34]

Obviously, the aforementioned complications with 1G biofuels in Africa do not encourage the channelling of entrepreneurship into the development of 2G biofuel capacity in the near term. As one interviewee noted, once 2G production is up and going, Brazil will be 'unbeatable', with Africa being once again the primary target for foreign production expansion plans, facilitated via ethanol diplomacy.[35] Yet, that time has not yet arrived and the current political challenges faced by the country suggest it might still be way into the future (see also UNCTAD 2016). Interviewees stressed that at such an early stage of 2G development, it is unlikely for technological breakthroughs and cost reductions to be achieved in Africa due to its poor infrastructure and lack of rules and finances: '*Africa lacks infrastructure, policies and skilled workers... it is not a good strategy to start with 2G in Africa as we firstly should understand more about the technology, problems and economic feasibility in Brazil*'.[36]

International organisations have stressed the need to avoid the emergence of technological gaps in this area between countries (see UNCTAD 2014). While actors like Brazil, China, the EU and the USA have all invested in 2G biofuels, developing countries at large have been absent from this technological race, which could lead to a growing disparity between country capacities. Support for remedial action has been proposed, including:

> [Promotion of] policies supportive of international joint ventures which can help provide access to intellectual property owned by international companies, as well as improve regulatory climates attractive of investment in poorer countries. By often having natural environments conducive for biomass production, developing country partners in international joint ventures might contribute host sites for demonstration and first commercial plants, as well as avenues for entering local biofuels markets. (UNCTAD 2014, p. 54)

Another option would be for Brazil to support some early 2G initiatives that have started to take place in Africa. For example, in November 2015, the South African

[33] Interview, Delegation of Brazil to the EU, Brussels, September 2013.

[34] Interview, BNDES, Sao Paulo, March 2013.

[35] Interview, BNDES, Sao Paulo, September 2014.

[36] Interview, Petrobras Biocombustível, Rio de Janeiro, March 2013.

government, through its Technology Innovation Agency (TIA), announced a call for proposals focusing on the demonstration of 2G and 3G technologies for biofuels production.[37] South African institutions, like Stellenbosch University, have already been active in this area. In particular, researchers there are doing fermentation work in the laboratory, up to 100 L scale (kilograms), using paper sludge, with fermentation residues going to biogas production. Furthermore, the first passenger flight with sustainable aviation biofuel was performed in 2016 from Johannesburg to Cape Town, as a result of a cooperative project between Boeing, South African Airways and other partners (see UNCTAD 2016; Boeing 2016).

While still a long way ahead, there are actions that Brazil can currently undertake in Africa in order to pave the way for a potential future introduction of 2G technology in Africa. After all, irrespective of the county setting, future 2G production will, in all places, build upon 1G infrastructure and technological know-how. Again, however, a cautious note is pertinent here. In particular, the possible impacts of pro-2G policies on food prices must be taken into account. It has been shown that even when 2G does not directly increase food prices, negative price effects might be generated depending on the type and quantity of land used, as well as on the impacts on other land uses (Thompson and Meyer 2013). Given the range of biofuel land-use and food-security-related concerns expressed worldwide, such reservations may severely constrain 2G production in Africa, especially if industrial development and deployment takes place in the absence of efforts and research to ensure their sustainability.

To conclude with Africa, very little progress has been made on the science production side. Technology transfer will simply not be enough, as it would be necessary to initiate research programmes involving African institutions and scientists from Brazil, the EU and the USA, so that African countries are given the opportunity to establish the most appropriate strategies for the production of ethanol in their countries. An effort has to be made to ensure that collaboration among all these actors does not suffer from the same disconnection problems described above for Brazil. However, internal political stability in Brazil is a *sine qua non* if the country is to effectively demonstrate its ability to act and influence externally.

11.4 Conclusion

This chapter has examined the current state of affairs with respect to the 2G sector in Brazil. Among others, it highlighted the lack of a clear strategy to connect fundamental science production with technological and industrial developments at the private and public sector levels. This has in turn compromised Brazil's potential to fully participate in the global race for 2G technologies. As a result, connections with the EU were limited to fundamental sciences and failed to influence private investment undertakings by domestic and external actors. With respect to the USA, no

[37] See: http://www.tia.org.za/Proposals

connections at the fundamental science level were established, with US companies relying on American technological know-how when commencing operations in the Brazilian market.

In addition, the opportunities and challenges for expanding 2G production and trade in Africa have been discussed. Growing investments in 2G R&D have taken place in Brazil during the past few years, stimulated largely by the financial support provided to domestic and international private companies by national agencies (i.e. BNDES and FINEP). Nevertheless, such strategies were almost entirely disconnected from the advances in fundamental science carried out by important projects such as the INCT-Bioethanol and SUNLIBB–CeProBio. Furthermore, Brazil's policy targets remained entirely focused on 1G consumption, showing a lack of concrete commitment by the Brazilian government when it came to 2G development. This contrasts with the approach adopted by major producers such as the USA, which has adopted an ambitious 2G consumption mandate through the 2007 Renewable Fuel Standard. The EU has also mobilised considerable finances and supported 2G in the 2009 RED, even if to a lesser extent than the USA. An increased level of policy coordination (i.e. with a transition from 1G to 2G) and budget support are essential for enabling Brazil's future leadership in the international 2G arena.

This chapter has also put forward a proposal for reviving Brazil's ethanol diplomacy, based this time on 2G and focusing on Africa as an example. Great potential for the expansion of Brazil's ethanol industry is found in the African continent. Although the establishment of a 2G market could deliver a range of socio-economic and environmental benefits to the local population, 2G development in Africa is not identified as a foreseeable option in the forthcoming years. The creation of a biofuels supply chain in Africa is constrained by a variety of factors, including high production costs and a lack of advanced technology needed for biofuel conversion, as well as the unattractive business environment to investors due to a lack of adequate legal and policy frameworks. Such barriers might affect negatively the opportunities for production generation within Africa. Yet, as this chapter has argued, Brazil has high potential, given that strategies are well adjusted, to pave the way for the future introduction of 2G capacity in Africa by working to mitigate the bottlenecks that have hindered the deployment of 1G biofuels. The creation of a 1G consumer market and establishment of a safe legal and policy environment are a *sine qua non* prerequisite for attracting prospective Brazilian 2G investments to Africa.

References

Afionis S, Stringer LC (2012) European Union leadership in biofuels regulation: Europe as a normative power? J Clean Prod 32:114–123

Afionis S, Stringer LC (2014) The environment as a strategic priority in the European Union-Brazil partnership: is the EU behaving as a normative power or soft imperialist? Int Environ Agreements Polit Law Econ 14(1):47–64

Afionis S, Stringer LC (2015) Promoting biofuels in developing countries through trilateral cooperation: Brazilian perceptions of the European Union and the United States in biofuels governance. SRI Working Paper No. 82. http://www.see.leeds.ac.uk/research/sri/working-papers/

Afionis S, Favretto N, Buckeridge M, Van Alstine J, Stringer LC (2013) The environment in Latin America and the Caribbean in the aftermath of the Rio+20 summit. In: Europa Publications (ed) South America, Central America and the Caribbean 2014. Routledge, London, pp 17–21

Afionis S, Stringer LC, Favretto N, Tomei J, Buckeridge M (2016) Unpacking Brazil's leadership in the global biofuels arena: Brazilian ethanol diplomacy in Africa. Global Environ Polit 16(3): 127–150

Baka J, Roland-Holst D (2009) Food or fuel? What European farmers can contribute to Europe's transport energy requirements and the Doha round. Energy Policy 37:2505–2513

Boeing (2014) Embraer and Boeing sign a memorandum of understanding for implementing a sustainable biofuels R&D center in Brazil. http://www.embraer.com/en-US/ImprensaEventos/Press-releases/noticias/Pages/Embraer-e-Boeing-assinam-Memorando-de-Entendimentos-para-implantar-centro-de-PD.aspx

Boeing (2016) Boeing, South African Airways and Mango Celebrate Africa's First Commercial Flights with Sustainable Aviation Biofuel on Boeing's 100th Anniversary. http://boeing.mediaroom.com/2016-07-15-Boeing-South-African-Airways-and-Mango-Celebrate-Africas-First-Commercial-Flights-with-Sustainable-Aviation-Biofuel-on-Boeings-100th-Anniversary

Cabral L, Shankland A, Favareto A, Vaz AC (2013) Brazil–Africa agricultural cooperation encounters: drivers, narratives and imaginaries of Africa and development. IDS Bull 44(4):53–68

Cotula L, Dyer N, Vermeulen S (2008) Fuelling exclusion?: the biofuels boom and poor people's access to land. IIED and FAO, London

Dalgaard KG (2012) The Energy Statecraft of Brazil: promoting biofuels as an instrument of Brazilian foreign policy, 2003–2010. PhD Thesis, Department of International Relations, London School of Economics

EIA (2013) International energy outlook 2013. U.S. Department of Energy, Washington, DC. http://www.eia.gov/forecasts/ieo/pdf/0484(2013).pdf

EPA (2015) EPA proposes renewable fuel standards for 2014, 2015, and 2016, and the biomass-based diesel volume for 2017. Office of Transportation and Air Quality, EPA-420-F-15-028, May 2015. https://www.epa.gov/sites/production/files/2015-08/documents/420f15028.pdf

European Commission. 2009. Directive 2009/28/EC of 23 April 2009 on the promotion of the use of energy from renewable sources. Official Journal of the European Union. L 140/16

European Commission. 2015. A framework strategy for a resilient energy union with a forward-looking climate change policy. COM(2015) 80 final

Favretto N, Stringer LC, Dougill AJ (2013) Unpacking livelihood challenges and opportunities in energy crop cultivation: perspectives on Jatropha curcas projects in Mali. Geogr J 180(4): 365–376

FGV (Fundação Getúlio Vargas) (2009). Feasibility analysis for biofuel production and investment recommendation. http://bibliotecadigital.fgv.br/dspace/bitstream/handle/10438/6944/328.pdf?sequence=1

Goldemberg J (2008) The challenge of biofuels. Energ Environ Sci 1(5):523–525

Government of Brazil (2013) Plano Decenal de Expansão de Energia 2022. Ministério de Minas e Energia. MME/EPE, Brasília

Government of the US (2007) Energy independence and security act of 2007. PL 110–140. http://www.gpo.gov/fdsys/pkg/BILLS-110hr6enr/pdf/BILLS-110hr6enr.pdf

Guardabassi P, Goldemberg J (2014) The prospects of first generation ethanol in developing countries. In: McCann MC, Buckeridge MS, Carpita NS (eds) Plants and bioenergy. Springer, New York, pp 3–11

Hamilton JD (2009) Understanding crude oil prices. Energy J 30(2):179–206

Havlík P, Schneider UA, Schmid E, Böttcher H, Fritz S, Skalský R, Aoki K, Cara S, Kindermann G, Kraxner F, Leduc S, McCallum I, Mosnier A, Sauer T, Obersteiner M (2011) Global land-use implications of first and second generation biofuel targets. Energy Policy 39(10):5690–5702

Hill J (2007) Environmental costs and benefits of transportation biofuel production from food- and lignocellulose-based energy crops. A review. Agron Sustain Dev 27:1–12
IEA (2011) Energy for all: financing access for the poor—special early excerpt of the World Energy Outlook 2011. OECD/IEA, Paris
IEA (2014) Renewable energy market analysis and forecasts to 2020: medium-term market report 2014. https://www.iea.org/Textbase/npsum/MTrenew2014sum.pdf
IPCC (2014) Climate change 2014: synthesis report. Contribution of Working Groups I, II and III to the fifth assessment report of the intergovernmental panel on climate change. IPCC, Geneva
Janssen R, Rutz DD (2012) Bioenergy for sustainable development in Africa. Springer, London
Khanna M (2008) Cellulosic biofuels: are they economically viable and environmentally sustainable? Choices 23(3):16–21
Köhler J, Walz R, Marscheder-Weidemann F, Thedieck B (2014) Lead markets in 2nd generation biofuels for aviation: a comparison of Germany, Brazil and the USA. Environ Innov Soc Trans 10:59–76
Luft G, Korin A (2009) Energy security challenges for the 21st century—a reference handbook. Preager Security International, Oxford
Lynd LR, Woods J (2011) Perspective: a new hope for Africa. Nature 474:20–21
Nonhebel S (2012) Global food supply and the impacts of increased use of biofuels. Energy 37(1):115–121
Sorrell S, Miller R, Bentley R, Speirs J (2010) Oil futures: a comparison of global supply forecasts. Energy Policy 38(9):4990–5003
SUNLIBB (2014) Sustainable Liquid Biofuels from Biomass Biorefining (SUNLIBB). Policy Brief #3—October 2014. http://www.york.ac.uk/org/cnap/SUNLIBB/pdfs/policy-brief-3-Oct2014.pdf
Thompson W, Meyer S (2013) Second generation biofuels and food crops: co-products or competitors? Glob Food Sec 2(2):89–96
UNCTAD (2014) The state of the biofuels market: regulatory, trade and development perspectives. http://unctad.org/en/pages/PublicationWebflyer.aspx?publicationid=1059
UNCTAD (2016) Second-Generation Biofuel Markets: State of Play, Trade and Developing Country Perspectives. http://unctad.org/en/pages/PublicationWebflyer.aspx?publicationid=1455
UNFCCC (2015) Federative Republic of Brazil intended nationally determined contribution towards achieving the objective of the United Nations Framework Convention on Climate Change. http://unfccc.int/focus/indc_portal/items/8766.php
USDOE (2008) DOE to invest up to $4.4 million in six innovative biofuels projects at US Universities. US Department of Energy Press Release. http://apps1.eere.energy.gov/news/news_detail.cfm/news_id=11974
Walter A, Segerstedt A (2012) International trade of biofuels: current trends and the potential role of Africa. In: Jannsen R, Rutz D (eds) Bioenergy for sustainable development in Africa. Springer, New York, pp 147–161

Index

M.S. Buckeridge, A.P. De Souza (eds.), *Advances of Basic Science for Second Generation Bioethanol from Sugarcane*, DOI 10.1007/978-3-319-49826-3

MIX
Papier aus verantwortungsvollen Quellen
Paper from responsible sources
FSC® C105338

If you have any concerns about our products,
you can contact us on
ProductSafety@springernature.com

In case Publisher is established outside the EU,
the EU authorized representative is:
Springer Nature Customer Service Center GmbH
Europaplatz 3, 69115 Heidelberg, Germany

Printed by Libri Plureos GmbH
in Hamburg, Germany